GEMSTONE

The Sitmar Liners
& the V Ships

1928 - 1998

The history of the shipping companies founded by
Alexandre and Boris Vlasov and of their vessels

by
Maurizio Eliseo

Carmania Press London
1998

*This book is dedicated to Gabriele and his parents Cristina and Vittorio, my great friends.
Welcome to Gabriele Magnaghi, who was born at the same time as this book.*

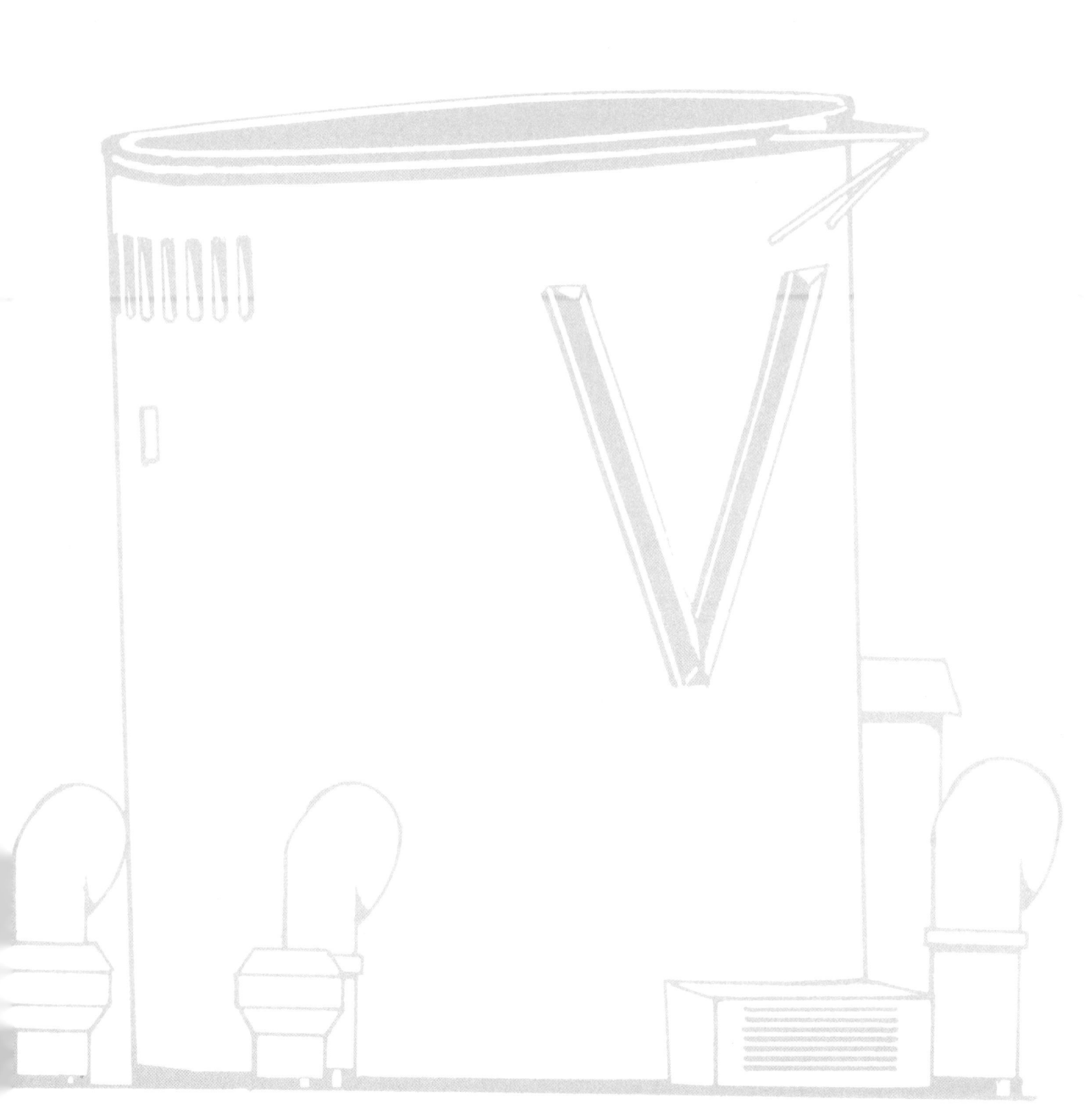

Foreword

by Mauro Terrevazzi, V.Ships Chairman

First of all I would like to express my deepest gratitude to the founder of the Vlasov Group, the late Alexandre and to his son Boris, who continued managing the company until his death in 1987.

The shipping division of the Vlasov Group has embraced a wide variety of operations in all corners of the World since its foundation.

This book is not only a meticulous record of the development of our Group and its personnel but can also be regarded as a chapter of maritime history over a period of major political and social changes and technological advancements in the design, construction, operation and management of shipping.

The author is to be congratulated on the quality and detail of the presentation which will be of absorbing interest to all those involved in the marine industry. Maurizio Eliseo has in my view done a good job to tell the story of the company with the pictures of vessels owned and operated over the years.

The hardware is important but the history of the Group has been made by thousands of men and women.

We enter with pride into the new millennium.

Mauro Terrevazzi
Principality of Monaco, November 1998

Above: the present V.Ships logo, introduced in 1984.
Below: V.Ships' chairman Mauro Terrevazzi has a particularly fond memory of the *Castel Felice* (here seen departing Southampton) as she was the first ship on which he worked; they both entered the company's service in 1952.
(Glasgow University Archives)

Introduction & Acknowledgement

In 1988, while researching the Italian liner *Rex*, I found in an old industrial photographer's shop in Genoa some dusty boxes of glass-plate negatives with magnificent shots of the great ship under construction. Needless to say, I ordered prints from all of them, but when I carefully inspected the envelope containing the prints a few weeks later, I was surprised to find among them a picture of a "tramp" with an aspect much different from the sublime elegance of the *Rex*. This rare photograph of the first *Fairsea* (published on page 74), portrayed as an emergency emigrant carrier, was my first encounter with a "V" ship.

Later, while researching Italian emigrant ships with Paolo Piccione for our common book projects (such as "The Costa Liners"), we amassed a certain quantity of material and photographs relating to Sitmar vessels. I became fascinated since then with the character of Alexandre Vlasov, his skill and ingenuity which led from his humble Cossack beginnings to his becoming one of the greatest shipowners in the World.

Paolo, as usual, put at my disposal his great maritime knowledge as well as his private library, which located on the top floor of a Twelfth Century tower in Genoa (Palazzo Salvago), I believe to be the largest existing collection of historical documents on Italian passenger ships.

Another providential coincidence while researching this book was that another great friend of mine, Roberto Fazi, with whom I had the privilege of working in Germany on the design and building of the *Costa Victoria* and *Costa Olympia*, moved in early 1997 to the Vlasov Group's newbuilding department together

The emigrant liner *Castel Verde* as portrayed in the first postcard published by Sitmar Line in 1950.

Introduction & Acknowledgements

with Arnold Brereton and Ken Norman, two authentic institutions in the passenger ship design field for the last twenty-five years or so. They unstintingly opened to me their files and memories and Roberto himself has been invaluable in introducing me to the whole staff (including the Vlasov family), locating information and documents, organising meetings and interviews with key figures of the company, either still in business or retired.

Researching and processing the data for this book has taken me some fourteen months and has involved journeys in Europe, Australia and United States. However, I would have taken much more time without the help of another friend, Alberto Bisagno, who is the expert on Sitmar Line "par excellence"; he sailed on Sitmar vessels for 15 years (also after the 1988 take-over by Princess Cruises) during which time, with the help of his brilliant wife Marina, he meticulously collected and filed vast amounts of data and papers on the company which he kindly put at my disposal.

Peter Vlasov and V.Ships chairman, Mauro Terrevazzi, who graciously agreed to write the foreword to this book, were a precious source of recollections and introduced me to others who were willing to help. Andrea Manzin, Boris Vlasov's nephew, with the help of his mother Olga Vlasov and his wife Roberta, was also a very important source of biographical notes on his ancestors.

Other people presently employed within the Vlasov Group deserve special mention: Tullio Biggi (V.Ships president), Ettore Bonaventura (Vlasov Shipholdings managing director), Roberto Giorgi (V.Ships managing

A tourist class baggage tag (from the collection of Arnold Kludas, Grünendeich) and a poster used in Australia to attract passengers to the *Fairsea*'s return voyages to Europe;
both date back to the 'fifties.

INTRODUCTION & ACKNOWLEDGEMENTS

director), Lorenzo Malvarosa (V.Ships technical operation director), Mario Bertolotti (fleet superintendent), Giorgio Matta (marine & planning manager), G.B. Arrigo (crew welfare manager and the Group's magazine editor), Sophie Bensaid (newbuilding secretary) and Per Bjornsen (project development manager), who kindly prepared the "V.Ships today and tomorrow" concluding chapter. Among the former managers and masters of the Vlasov Group I had the good fortune to interview Ardavast Postoyan, employed with Vlasov since 1938 and later Boris' right-hand man, Giorgio Lauro, Sitmar Line and Sitmar Cruises director in crucial years, Matteo Parodi and Dario Rivera, technical managers during the earlier conversions, Capt. Rodolfo Potenzoni and Capt. Ferruccio Rocconi, masters on almost all Sitmar passenger ships.

I also acknowledge the kind support of Elsa Hebe Lleonart, Alexandre Vlasov's secretary during the War years and today's vice-president of Canumar (the Argentine company he founded in 1941), and Adelaide Planchette, Boris Vlasov's secretary; both of them obviously had a privileged observation point of the Vlasov Group history and graciously shared their first-hand recollections with me.

Another three active researchers and historical advisors, whose dedication to this project was admirable, must be mentioned for devoting a lot of their time at libraries finding original papers and pieces of information: Admeto Verde from Naples at Lloyd's Register of Shipping, Sitmar enthusiast Gerald Laver from Leongatha, Australia and Anthony Cooke from London to whom, incidentally, goes also my sincere gratitude for supporting and publishing this work.

This book benefits also from some masterpieces of technical drawing by a friend who gave me months of his talented hand, Enrico Repetto from Genoa, industrial designer and professional draughtsman. He meticulously produced the profiles herewith published and in his drawings it is difficult to find a single porthole or a stanchion which is out of place. Enrico has given me the privilege of publishing for the first time ever accurate side elevations of ships such as the *Rio de la Plata*, the *Oxfordshire*, the *Castel Felice*, the Victories and many others...

And another great friend, the world-renowned marine artist Stephen J. Card, graciously agreed to realise the cover painting of *Castel Felice*: looking at it you can

Above: a tea cup and a coffee cup used in first class and tourist class, respectively in the 'fifties and in the 'sixties, before Sitmar Line was transformed into Sitmar Cruises.

INTRODUCTION & ACKNOWLEDGEMENTS

almost hear the sound of the waves and breath the sea breeze… The beauty of his oil on canvas speaks by itself…

Furthermore, my sincere "grazie" go to Francesco Dell'Olio, Giuseppe Dell'Olio, Stefania Bosi, Giorgia Matarese and Elda Genovese at GDL Comunicazione, Genoa, for designing and following step by step the art production of this book. Other people who had an important role in the completion of "The Sitmar Liners & the V Ships" and to whom I express a word of sincere thanks are, in alphabetical order: Declan Barriskill at Guildhall Library (London), Julie Benson at Princess Cruises (Los Angeles), Jonathan Boonzaier (Singapore), Nello Brancaccio (Genoa), Ian Carter at the Imperial War Museum (London), Anne Cowne at Lloyd's Register of Shipping (London), Arthur W. Crook (Forest Row), the Dinav, Dipartimento di Ingegneria Navale (Genoa), Alex Duncan (Gravesend), Laurence Dunn (Gravesend), Ann Einsiedler at the Navy Museum (Washington D.C.), Kathy Flynn at the Peabody Essex Museum (Salem), George Gardner at the University of Glasgow, Lyn Gardner at the Mariners' Museum (Newport News), Mark Goldberg (Baltimore), Jacqueline Gooze at Vickers PLC (London), Ambrose Greenway (London), Andres Hernandez (Miami), Mary E. Hoban at the Hartlepool Borough Council, David Hodge at the National Maritime Museum (Greenwich), Karen Howatt at Govan Workspace Ltd (Glasgow), Dick Keys (Newcastle-upon-Tyne), A. M. Jackson at the Glasgow City Council, Walter W. Jaffee (Benicia), Helmut Kannenberg (Odessa), Andrew Kilk (Oakland), Arnold Kludas (Grünendeich), Peter C. Kohler (Washington D.C.), Hans-Peter Lemcke-Braselmann (Wuppertal), Jan J. Loeff (Fort Lauderdale), Mike Macdonald (Brussels), Moira MacKay at University of Glasgow, Luciano Mangini at Rina, Registro Italiano Navale (Genoa), Tom McCluskie at Harland & Wolff (Belfast), William H. Miller (Secaucus), Giovanna Monti at Lufthansa (Milan), Peter Newall (Blandford Forum), Mark Nicholl at Cambridge University Library, Hisashi Noma (Tokyo), Jim Nurse (London), Maria Enrica and Paolo Oblieght (Monaco), Lyn Palmer at P&O (London), Achille Rastelli (Milan), Paolo and Piero Rigo at Associazione Marinara Aldebaran (Trieste), Mario Sannino (Genoa), the late Len Sawyer (London), William Schell (Holbrook), Antonio Scrimali

Above: the V.Ships coffee cup presently in use and the tea cup, sporting the Sitmar Cruises stylish logo with the crossed initials of the company, introduced in the early 'seventies for stateroom service on board the sisterships *Fairsea* and *Fairwind*.

INTRODUCTION & ACKNOWLEDGEMENTS

The new corporate colour scheme and logo designed in 1988 for the newbuilding Sitmar FairMajesty. *(Alberto Bisagno, Genoa)*

(Alpignano), Flavio Serafini (Imperia), M. Smith at Vickers Shipbuilding & Engineering Ltd (Barrow-in-Furness), Dario Spigno at Rina, Registro Italiano Navale (Genoa), Ian Stewart (Rockingham Beach), R. G. Todd at the National Maritime Museum (Greenwich), Gordon Turner (Toronto), Paolo Valenti at Associazione Marinara Aldebaran (Trieste), Jan Van Der Elst (Heemstedt), Ian Whitehead at the Hartlepool Borough Council, Victor Young (Wellington), Giangiacomo Zino at T. Mariotti (Genoa).

Maurizio Eliseo
Principality of Monaco, October 1998

CHRONOLOGY

1880: Alexandre Vlasov was born of a family of Cossack peasants in Novocerkassk, a town 50 km north of Rostov, on the Don River.

1893: with the help of his grandfather he moved to Odessa; here he took several casual jobs and in the meantime he studied, graduating in Civil Engineering.

1910: after obtaining a job with the Odessa Municipality, he met and married a Russian girl, Vera. Employed by the Sanitary Department, he was in charge of inspections on board vessels calling at Odessa.

1913: Boris Vlasov was born on 13th March.

1917: the Bolshevik Revolution transformed Odessa (which was a stronghold of Bolshevik opposition since the famous 1905 mutiny of the *Potëmkin*), into a battlefield: strikes, pillage, mutinies, storming and occupation of public offices, particularly the Municipality where Alexandre worked. Being a representative of the detested Imperial government he was compelled to flee abruptly with his 4-year old son Boris to Kishinev and later to Bucharest. In the confusion he lost trace of his wife Vera, who he would meet again only many years later, after he had already married his second wife, Emilia Cavura, a widow with a daughter, Nadin.

1925: Alexandre Vlasov became general agent for Romania of Skarboferm, a firm which controlled coal mines in Poland. On behalf of many industrial and electrical companies he chartered vessels for the carriage of Skarboferm coal to Romania. Also, in this year, he purchased a share in the Romania Prima Societate Nationala de Navigatione.

1928: in April, Alexandre bought a small vessel for a single round-trip voyage between Danzig, Kiel and Kotka with a cargo of coal. This Romanian flag vessel was re-named *Boris* for the voyage in honour of his son. Her log abstract shows that the vessel cleared Danzig on 1st May, arrived Kiel on 3rd May, on to Kotka on 15th June and finally returned to Danzig on 7th August. This reliable information comes from the voyage records which are held at Lloyds in London. Unfortunately it has not so far been possible to trace the origin of the first *Boris* or her particulars.

1933: in February Alexandre Vlasov became Skarboferm General Agent in Italy and Greece. He bought, in partnership with the Greek shipowner P. Argyropoulos, a half share in his second vessel, the *Mimoza*, appointing C. Arvanitides, Piraeus, manager; he also took over financial control of Romania

Alexandre Vlasov, the Group's founder, photographed in 1947 together with his wife Emilia Cavura in his Argentine estate Dos Ríos.

Prima Societate Nationala de Navigatione: the Vlasov Group was born.

In November Alexandre founded in Athens (Commercial Bank Building, Odos Omirou 16) Scomar, Societé Commerciale et d'Armement, buying the 50% share of Mr Argyropoulos in the *Mimoza* (re-named *Mimosa*) and purchasing for the company another two vessels, the *Nadin* and the *Boris*. Incidentally, Nadin was the daughter of his wife, born from her first marriage, and she eventually became wife to his son Boris.

1934: the Romania Prima Societate Nationala de Navigatione was put into liquidation and its older steamers were sold while two vessels, the *Prahova* and the *Siretul* were transferred in September to Alexandre Vlasov's Societate Anonima de Navigatione, headquartered in 47, Strada Akademiei, Bucharest.

During the Fall of the same year Alexandre Vlasov opened a new company in Milan involved in the coal trade, Sitcom, Sindacato Italiano Combustibili S.A.

1936: in September Alexandre founded in London the Campden Hill Steamship Co. Ltd, in partnership with the well-known Greek shipowner Manuel Kulukundis (his descendants are still in the shipping business at the time of writing); they bought an old steamer of the standard War class, entrusting her management to Counties Ship Management Co. Ltd (headquartered at Bury Court House, London); their ships all had names ending in "Hill" and Vlasov's vessel was given the name *Campden Hill*.

1937: Alexandre opened two other companies in London, Alva Steamship Co. Ltd, shipowners, and Navigation & Coal Trade Ltd, shipbrokers and shipmanagers, with headquarters in Billiter Buildings.

He bought a vessel under construction in Sunderland, which was launched on 30th December as *Gemstone* (starting the tradition of the "stone" class of freighters) and he ordered two similar vessels, delivered in 1938 as *Starstone* and *Lodestone*. These were the first newbuildings in the history of the Group.

Boris Vlasov and his wife Nadin, portrayed in St. Moritz during Winter 1938/'39 soon after they married.

Chronology

1938: with the sudden and bloody ascent to the Romanian throne of the dictator Carol II, Alexandre Vlasov closed down his business in the country and moved with his family to Milan.
Boris Vlasov graduated in High Tension Electrical Engineering at the Vienna Polytechnic and married Nadin; he received as a "gift" from his father the nomination to president of his newly established Sitmar Line. Boris and Nadin would have two daughters (Olga in 1939 and Tania in 1941) and a son, Alexander, in 1948.
On 30th April Alexandre founded Sitmar, Società Italiana Trasporti Marittimi S.A., with a deed by the notary Virgilio Neri of Milan and with an initial capital of 1000 shares of 1000 Italian Liras. The headquarters were established in the same office as Vlasov's Sitcom, at Via del Conservatorio 15 in Milan, with a small branch office in Genoa (after the War, Sitmar's headquarters were moved in Via Santa Tecla, 5).
Alexandre Vlasov appointed Carlo di Stefano and Boris Demcenko general managers of the new Sitmar and Luigi Valazzi managing director. The latter was an old friend of the Vlasov family; they met for the first time when they were living in Odessa. At the time Alexandre Vlasov was sanitary inspector of the local port authority and Valazzi's father was the general manager of the Odessa office of the old Italian Sitmar Line (Società Italiana Servizi Marittimi S.A.). This company had been absorbed by Lloyd Triestino and closed down in December 1931 and now it was decided to revive the name, very famous in Mediterranean ports thanks to the company's beautiful vessels, as a good omen for Vlasov's new shipping concern. However, this sometimes caused confusion because three other shipping companies, Sitmar (Sbarchi, Imbarchi & Trasporti Marittimi S.A.), Trasporti Marittimi S.A. and Citmar (Compagnia Italiana Trasporti Marittimi S.A.) were operating at the time in Italy.

1940: with Italy prepared to enter the War against the Allies, Alexandre Vlasov moved with his family first to Lausanne in Switzerland, and after a few months moved briefly to New York. Here he founded the Alvion Steam Ship Corporation Inc. and its Panamanian subsidiary, the Dolphin Steam Ship Corp.

1941: in February Alexandre settled in Buenos Aires and acquired Argentine citizenship. He purchased a large estate near Cordoba, "Dos Rìos" ("Two Rivers"), now administrated by his descendants.
On 1st August he opened in Buenos Aires (596, 25 de Mayo Street) the Compañía Argentína de Navegácion de Ultramar S.A. (better known by the acronym Canumar), transferring to the new company two of his Romanian vessels, the *Prahova* and the *Oltul*. He engaged a young and brilliant secretary, Mrs Elsa Hebe Lleonart, who is nowadays the vice-president of Canumar.

Artist's impressions of life on board the *Fairsea* from a 1954 brochure realised by A. Storace.

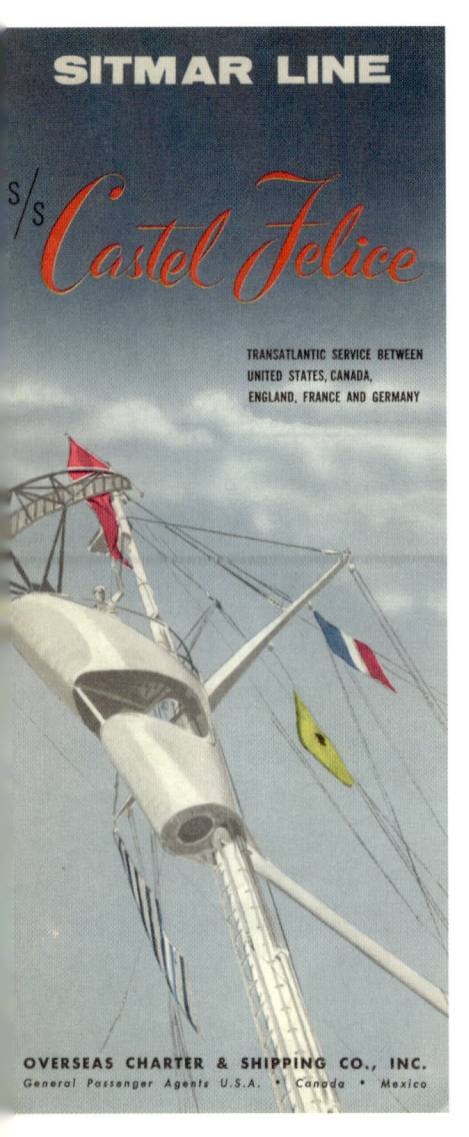

CHRONOLOGY

1943: on 20th August Alexandre founded in Buenos Aires another shipping concern, the Compañia Sud Americana De Export & Import, Soc. de Resp. Limitada (Cosadex S.R.L.) for the management of two small fishing trawlers owned by Canumar, the *Besugo* and the *Lenguado*, later joined by a similar vessel, the *Corvina*. Cosadex was closed down on 17th August 1967.

1947: on 21st July, in Geneva, the International Refugee Organisation (I.R.O.) started its programme of mass resettlement in the Americas and Australia of new settlers and displaced persons, chartering passenger ships from private shipowners. Among them there was Alexandre Vlasov who bought for the purpose three standard American vessels, transforming them into the emigrant ships *Castelbianco*, *Wooster Victory* (later *Castel Verde*) and *Fairsea*.

Alexandre Vlasov, who was himself an emigrant and left his home at the age of 14 with "...only the pockets to put my hands" (as he used to say), felt strongly that the voyage should be as comfortable as possible for his "assisted passengers" (he prohibited the use in his offices of words such as emigrants or displaced persons) and thus he enrolled some famous managers of the prestigious Italian Line to work alongside him, such as Italo Verrando (former President of the Italian Line Inc. of New York), G.B. Ratto (former Corporate Chef de Cuisine of the Italian Line), passenger ship-designer Francesco Bruzzo (former Lloyd Sabaudo owner's superintendent during the building of the *Conte di Savoia*) and Alfredo Cappagli in running the passenger services on board his ships.

In November Boris moved with his wife and two daughters to Buenos Aires, also taking up Argentine citizenship.

1948: On 19th October the *Castelbianco* (ex-*Vassar Victory*) opened the passenger service of Sitmar Line clearing Genoa for Australia, fully booked with assisted passengers, under charter to the International Refugee Organisation.

1949: Alexandre Vlasov opened in Broadway, New York, a new company, Navcot Corp. Inc., headquartered in Bowling Green Building until June 1977 (when they moved to 130, Liberty Street), to which he transferred the ownership of the *Fairsea*, *Coralstone* and *Rubystone*.

1950: in May the *Wooster Victory* was re-named *Castel Verde* and transferred to the new Vlasov South American line.

The Vlasov Group entered the oil market with an impressive newbuilding programme of six tankers; two of them were ordered from John Brown & Co. Ltd (*Almak* and *Algol*), another two from Nederlandsche Dok en Scheepbouw (*Alkaid* and *Alkor*), one from Sir James Laing and Sons Ltd (*Alva Star*) and one from Greenock Drydock Co. (*Alva Cape*). They were all to be delivered between 1952 and 1953.

This and following page: some colourful advertising material published by Sitmar Line in the 'fifties and 'sixties for their regular line voyages.

CHRONOLOGY

1952: Alexandre Vlasov settled in Genoa, joined two years later by Boris; after living for a few months in a suite at the Colombia Hotel, he bought a castle in Pieve Ligure, East of Genoa.
Navcot Australia Pty Ltd was opened in Margaret Street, Sydney, in a tiny former barber's shop. To manage the new premises George Grieshaber was recruited, a German émigré who in 1948 was the Australian liaison officer for the International Refugee Organisation. Mr "Sitmar" (as Grieshaber was dubbed in Australia) did a great job, using all his knowledge as a former travel agent, to establish Sitmar's reputation in Australia, enabling them to fill their ships north-bound with young "aussies" and disillusioned migrants looking for a passage back to Europe cheaper than that offered by the conference lines' official ships.
1953: in March the *Castel Bianco* was transferred to the new Vlasov Central America line and was later joined by *Castel Felice* and *Castel Verde*.
1954: Bellaccini and De Longis wrote the music and the words, respectively of "Sitmar... Castelli sul mar", official song of the Line.
1955: Sitmar was awarded the contract for the transport of assisted passengers by the Australian government.
1961: on 3rd October Alexandre died in Genoa, and Boris took full control of the Vlasov empire. He was very keenly involved in the design of the newbuild-

CHRONOLOGY

ings, especially the passenger ships, and he was particularly clever in dealing with all the technical aspects of shipping. For this reason he often preferred to leave the running of economic and administrative matters to his right-hand man, Ardavast Postoyan, still nowadays an institution at Vlasov Group headquarters, who had joined the company in 1938.

1962: on 2nd January the Vlasov Group headquarters were moved from Genoa to the Principality of Monaco, where they are presently located.

1964: in December the *Castel Felice* left Melbourne on the first cruise in Sitmar history. It was a 10-day Christmas and New Year jaunt to Noumea, Cairns, Hayman Island and the East Coast.

1968: Boris Vlasov had his fourth child, Peter, from his partner Annelore Zulke, whom he later married in 1985. To honour his Russian origins he gave to his four sons the names of the Tsar's children.

1969: Vlasov obtained the Italian state subsidy and the contract for the regular line service between Italy, Central America and the North Pacific previously run by the Italian company Italnavi (sold at the time to the Costa Line). To maintain the Savona-Vancouver service, Vlasov founded the Ital Pacific Line, ordering four freighters from a German shipyard for his new concern.

1971: with Sitmar's loss of the Australian migrant contract to Chandris Line and the introduction of the new *Fairsea* and *Fairwind*, Sitmar Line abandoned regular passenger services and was transformed into Sitmar Cruises Inc. with headquarters in Los Angeles. Boris established a Trust to which all his companies' assets were transferred in order to ensure the continued economic stability of the

Below, left: one of the certificates given to passengers to mark the crossing the line ceremony.
(Alberto Bisagno, Genoa)

Group after his death. In August Sitmar, together with Shaw Savill Line, established a net of tourist offices in major European, Australian and New Zealand cities, called Sea Travel Centres, to offer a modern system of marketing and sales arrangements for their cruise product.

1973: in November Navcot Shipping Holdings Ltd, a joint-venture between Vlasov Group and Capitalfin (a financial company controlled by the Italian Banca Nazionale del Lavoro, Fiat and Montedison), bought Shipping Industrial Holdings Ltd, the second largest shipping group in the United Kingdom and parent company to two prestigious British shipping firms: Dene Shipping Co. Ltd (founded in 1937) and Silver Line (founded in 1925).

1975: the first Western shipping concern in Saudi Arabia, Amar Line of Jeddah, was jointly-founded by Vlasov and Gaith Pharaon to run the freight and oil route between Middle East and Mediterranean ports. Thanks to this move the Vlasov Group could survive the oil crisis which caused a crippling increase in fuel oil costs. The new company took over three freighters and two tankers owned by the Vlasov Group. Its subsidiary, Saudi Maritime Transport Co., started a regular passenger and car ferry route between Jeddah and Suez.

1976: the former newsletter published by Silver Line became the official magazine of the Vlasov Group. This publication had started in 1964 as a modest duplicated sheet recording company events and became a properly printed magazine in 1971. In March 1976 the Vlasov banner appeared on issue number 74, with a main article by Silver Line's chairman Robert G. Crawford significantly titled "Jointly we survive". Since then this newsletter, which emphasises the Group ship operations and those people directly concerned with them, has been printed and distributed to all the Group's employees. Now called V. Ships News, it represents an important means of communication and human contact in a large, International shipping concern.

1977: the Vlasov Group bought out the whole shares of Capitalfin, becoming the sole owner of Shipping Industrial Holdings Ltd, which was merged with Navigation & Coal Trade Co. Ltd, Silver Line becoming their trademark.
In May Tradax entrusted Shipping Management S.A.M. with the running of their Panamax OBOs *Carisle* and *Carlantic* and their bulk carriers *Amsterdam* and *St. Nazaire*. Tradax was the first customer of Vlasov's Shipping Management S.A.M. This important achievement, which started what is now the main activity of the Group, happened after the *Arapaho*, the former Silver Line *Seto Bridge*, had been bought by Tradax from the Vlasov Group itself the previous year, with the management remaining in the hands of the latter.

1979: in March, in a major move designed to strenghten Sitmar's position in the new cruise market which had replaced the traditional liner voyages, a brand new

Previous page, above; Monaco Monte-Carlo, 1968: Boris Vlasov with his nephew Andrea Manzin, wearing the T-shirt of the family yacht Shark. In March 1972 a new Shark (previous page, below), was delivered to Boris by Abeking & Rassmussen of Lemwerder; with a gross tonnage of 330 and an overall length of 43 metres, she was driven by a couple of 2100 HP Caterpillar engines.
(Andrew Kilk, Oakland)

CHRONOLOGY

The Sitmar liners, such as the first Fairsea *here portrayed, used Rome as port of registry. Note the fine stern shape of the C3 standard vessels. (Arnold Kludas, Grünendeich)*

cruise marketing office was opened in Sydney. Ted Blaimed was appointed managing director while the other department executives were Capt. Luigi Nappa, Geoffrey Starkie, Phil Young, Bruce Gregory and Thomas Cook.

On 31st May in Ulsan, Korea, the 30,700 dwt bulk carrier *Docegulf* was delivered to the Brazilian company Docenave. The Vlasov newbuilding team played an important role, for the first time on a non-Vlasov ship, during the design and building stages of the vessel. Shipping Management S.A.M. was then entrusted with the running of the ship and later of all her fleetmates. Thus, after Tradax, Docenave became an important V.Ships customer.

1981: in November the contract was signed for the building of the first brand new cruise ship of Vlasov, the fourth *Fairsky*. The newbuildings programme continued with the deal for the construction of three more larger cruise ships, the *Sitmar FairMajesty* and the eventual *Crown Princess* and *Regal Princess* of Princess Cruises.

1983: with Costa Line of Genoa in financial troubles, Vlasov made a serious proposal to buy the *Eugenio C.* and transform her into a full time cruise ship to work alongside their *Fairstar* in Australian waters. Eventually, in October, Costa Line was able to sell the elder *Federico C.* and thus the *Eugenio C.* was withdrawn from the sale list.

1984: in January all the Group's shipping management skills were incorporated under the name V.Ships, with the incomes of the company shared at 50% between the Vlasov family and its senior managers. For the occasion a new flag logo was designed to replace the old one (see page 5).

The Swedish Johnson Line became the third large customer of V.Ships, after Tradax and Vale Do Rio Doce Navegaçao (Docenave Line).

Silver Line was re-organised as a commercial management company in the chartering, sale and purchase field.

1986: in December the vessels managed by V.Ships had already reached a total of eighty.

1987: in September the new Sitmar logo and hull colour scheme, designed for the newbuilding *Sitmar FairMajesty* (see page 10), was applied to *Fairwind*, which had her registered name altered to *Sitmar Fairwind*.

On 2nd November Boris Vlasov died in Kumamoto, Japan.

1988: Sitmar Cruises was put up for sale by the Trust formed by Boris himself in 1971, as an economic safeguard for his family. On 28th July the company was sold to P&O. They assigned all Sitmar's ships to Princess Cruises, exception made for *Fairstar*, for which a separate management office was kept in Sydney. All other vessels were re-christened with Princess names. To the famous British shipping firm went as well the *Sitmar FairMajesty* (at the time in an advanced stage of fitting out), and the Italian company Astramar S.p.A. of Palermo, which owned the building contract of two 70.000 tons cruise vessels under construction in the Fincantieri shipyard in Trieste, the eventual *Crown Princess* and *Regal Princess*, delivered respectively in 1990 and 1991.

1991: in January was formed a new company called Vlasco, a 50/50 joint venture between Black Sea Shipping Co. (Blasco) and Vlasov. Aim of the new company was crew recruitment and training in Ukraina, and management of 60 multi-purpose vessels owned by Blasco and all registered in Ukraina. Blasco is one of the World's largest ship owning companies.

1993: on 18th June the ex-*Fairwind*, bearing at the time the name of *Dawn Princess*, was sold back by P&O to the Vlasov Group, becoming the first passenger ship owned by the company since the 1988 sale of Sitmar to P&O itself.

1994: to give access to fresh capital and new investment power, 20% of V.Ships

The *Fairsky*, although built from a C3 hull identical to the *Fairsea*'s, had a more modern and sophisticated appearance and interior accommodation of a higher standard: for her 1958 conversion, in fact, Vlasov called for the first time on an architect to design her interior décor.
(Ambrose Greenway, London)

CHRONOLOGY

was sold to G.E. Capital, the financial services arm of General Electric Corp.
In April the *Silver Cloud* entered service, followed in January by her sister ship *Silver Wind*. They probably are the best recent example of full concept management offered by V.Ships to external customers. V.Ships organised for the owner the foundation of the company, the market strategy and target and the complete design of the five-star vessels, from the drawing board to the in-service management.

1996: the *Minerva*, delivered in April, was the first passenger ship in a series of newbuildings which re-introduced Vlasov into the cruise market.

1998: a date of important anniversaries. 70 years ago Alexandre Vlasov started his shipping activities; 60 years ago Sitmar was founded and 10 years ago was sold to P&O.

Today the Vlasov Group is one of the biggest shipping concerns in the World. Its shipping management arm, incorporated under the trademark V.Ships, manages over 400 vessels for almost 80 International customers (including 25 ships owned by the Liberian Vlasov Shipholdings Inc.) through the Monaco headquarters and 25 branch offices (for further information see the chapter "V.Ships today & tomorrow", page 194).

To commemorate this important anniversary a museum dedicated to the shipping activities of Alexandre and Boris Vlasov and their vessels is being organised at the Odessa Maritime Museum, where everything started.

And this book has been published as a tribute to two of the greatest shipowners of this Century and to preserve the memory of all the other people and of the vessels which enabled the Vlasov Group to enter the 21st Century so confidently.

The bust of the Group's founder, specially realised for the permanent pavillion at the Odessa Maritime Museum; it will be dedicated to the history of the shipping companies founded by Alexandre and Boris Vlasov and of their vessels.

CHRONOLOGY

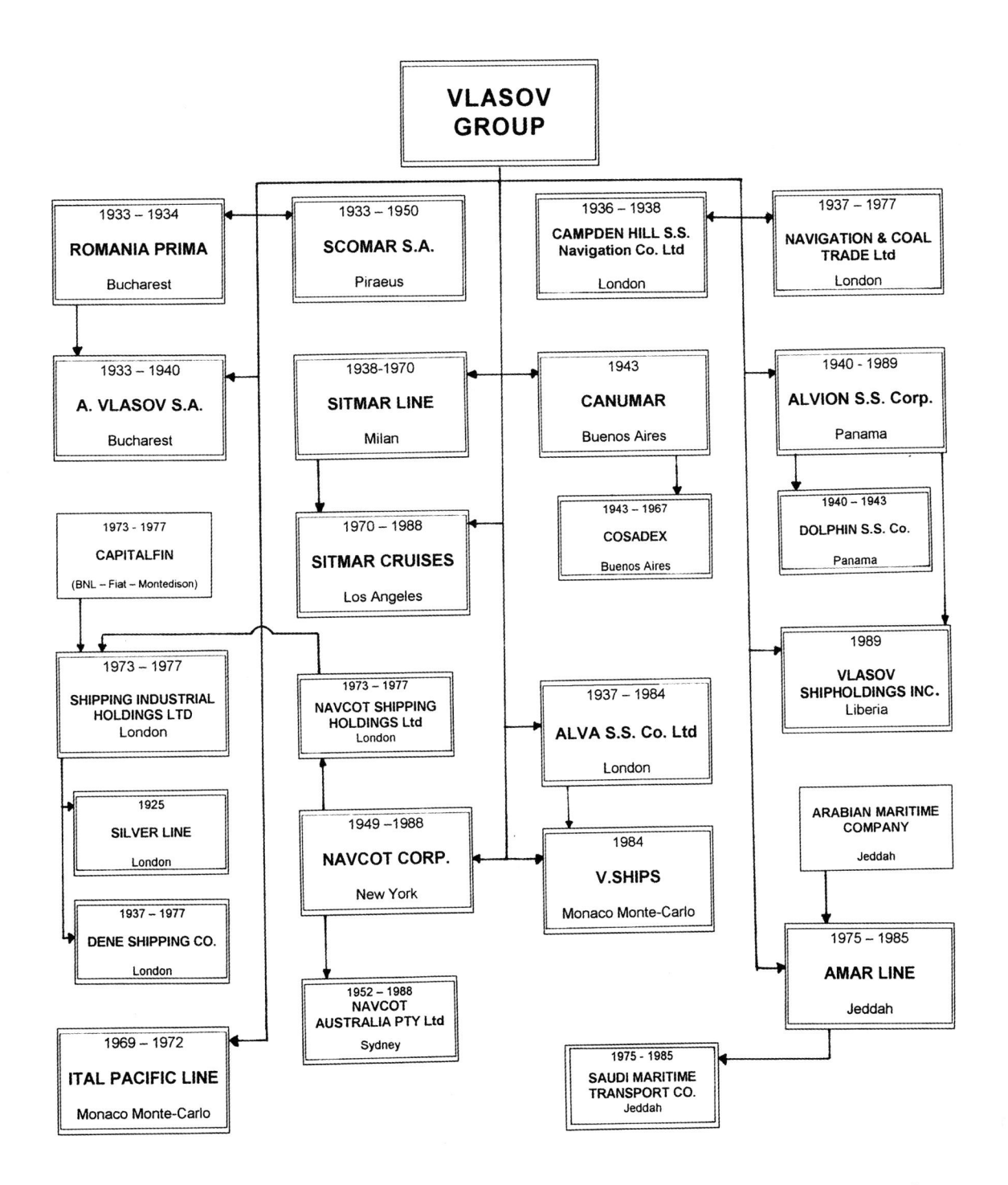

```
                            VLASOV
                            GROUP

1933-1934      1933-1950        1936-1938             1937-1977
ROMANIA PRIMA  SCOMAR S.A.      CAMPDEN HILL S.S.     NAVIGATION & COAL
                                Navigation Co. Ltd    TRADE Ltd
Bucharest      Piraeus          London                London

1933-1940      1938-1970        1943                  1940-1989
A. VLASOV S.A. SITMAR LINE      CANUMAR               ALVION S.S. Corp.
Bucharest      Milan            Buenos Aires          Panama

                                1943-1967             1940-1943
                                COSADEX               DOLPHIN S.S. Co.
                                Buenos Aires          Panama

1973-1977      1970-1988                              1989
CAPITALFIN     SITMAR CRUISES                         VLASOV
(BNL-Fiat-                                            SHIPHOLDINGS INC.
Montedison)    Los Angeles                            Liberia

1973-1977      1973-1977        1937-1984
SHIPPING       NAVCOT SHIPPING  ALVA S.S. Co. Ltd
INDUSTRIAL     HOLDINGS Ltd
HOLDINGS LTD
London         London           London

1925           1949-1988        1984                  ARABIAN MARITIME
SILVER LINE    NAVCOT CORP.     V.SHIPS               COMPANY
London         New York         Monaco Monte-Carlo    Jeddah

1937-1977                                             1975-1985
DENE SHIPPING CO.                                     AMAR LINE
London                                                Jeddah

               1952-1988
               NAVCOT
               AUSTRALIA PTY Ltd
               Sydney

1969-1972                       1975-1985
ITAL PACIFIC LINE               SAUDI MARITIME
                                TRANSPORT CO.
Monaco Monte-Carlo              Jeddah
```

Diagram of the main companies founded by Alexandre and Boris Vlasov from the 1933 foundation of the Vlasov Group up to now. The years over the subsidiary's name refer to the dates when it was opened and, eventually, closed down or incorporated under a new name; in the latter case an arrow unites the boxes containing the different names of the same company. The town under the company's name indicates the location of the headquarters. Subsidiaries of a main company are indicated by a smaller box. The boxes with a single line border indicate companies outside the Vlasov Group which shared with it the ownership of a subsidiary.

Freighters and bulk carriers

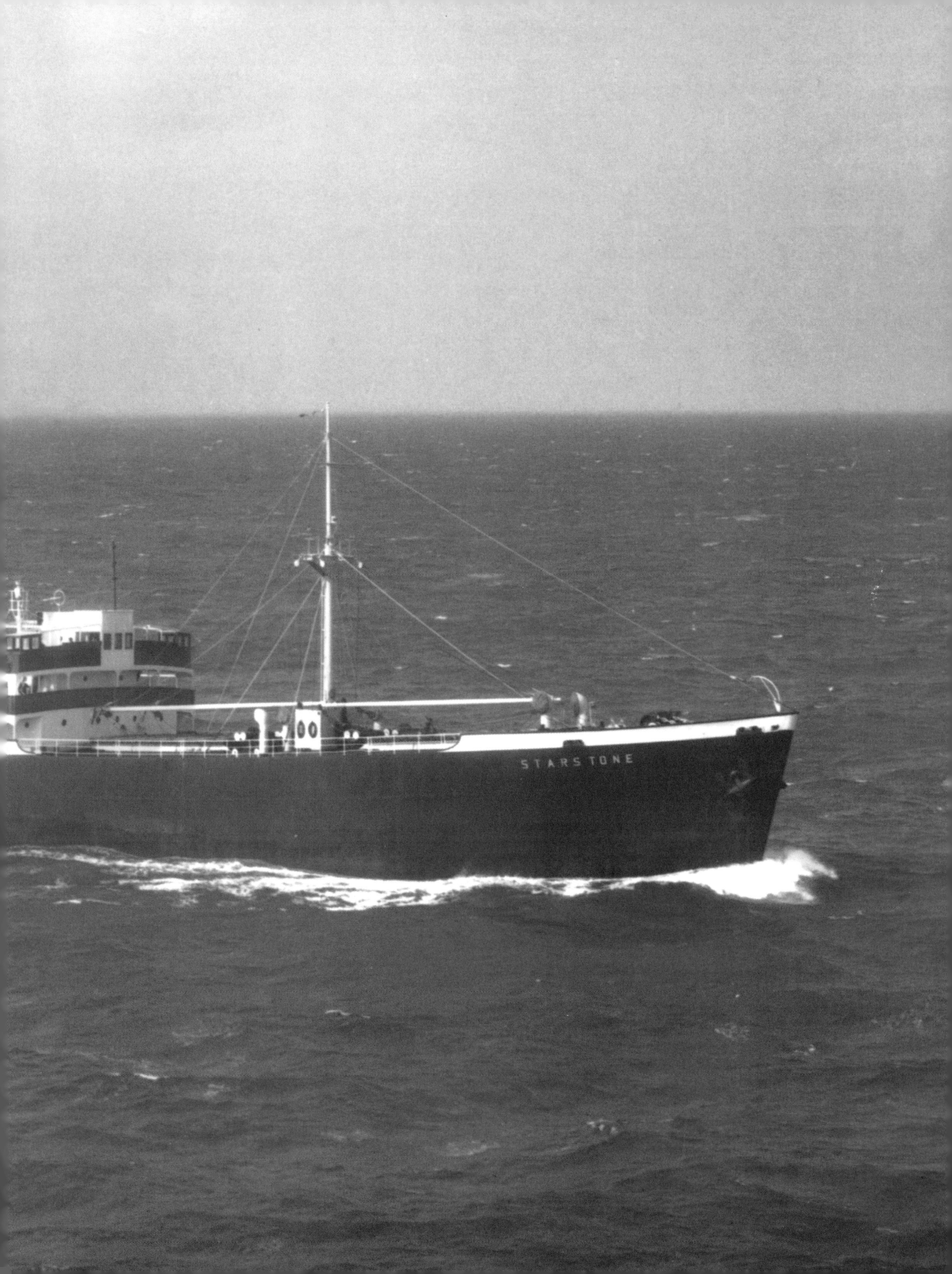
STARSTONE

Alexandre Vlasov became a "de facto" shipowner in 1933 when he took over financial control of a Romanian company and founded a new concern in Greece. However his activities in the shipping field can be traced back to 1928, when he bought a small Romanian trawler of 150 nrt, naming it after his son Boris. The *Boris* was employed for a few months in the Baltic coal trades.

Little information about her career can be found in the voyage records of Lloyd's of London, where the ship was recorded for insurance purposes from 22nd April 1928.

In September 1936 Alexandre Vlasov and the well-known Greek entrepreneur Manuel Kulukundis jointly founded the first British-registered firm in the history of the Group, the Campden Hill Steamship Co. Ltd, named after its sole vessel, *Campden Hill*. She was joined one month later by two steamers which introduced the tradition of giving the Vlasov's freighters names ending in "stone", the *Sunstone* and *Pearlstone*.

The joint-venture with Kulukundis was however a brief one. In 1938 Alexandre Vlasov was financially strong enough to start his own shipping company in England, the Alva S.S. Co. Ltd. A second company, the Navigation & Coal Trade Co. Ltd was formed to act as shipmanagers; thus the Campden Hill S.S. Co. Ltd was put into liquidation.

Alva commenced activities by purchasing the building contract for a steamer already under construction at the yard of James Laing & Sons in Sunderland for the Minster Steamship Co. Ltd of London. The vessel was launched on 30th December 1937 as the *Gemstone*, becoming the first brand-new vessel in the Group's history. Two similar vessels, the *Starstone* and *Lodestone*, left the builder's slipways a few months later.

In April 1938 Vlasov founded in Milan the Sitmar Line, using as headquarters the same office as Sitcom, Sindacato Italiano Combustibili S.A., which had been opened by Vlasov in 1934 to manage the importation of coal into Italy. To provide an instantaneous fleet for the new Italian concern, the former Campden Hill vessels *Pearlstone* and *Sunstone* were transferred to Sitmar as *Castelnuovo* (soon renamed *Castelbianco*) and *Castelverde* respectively.

The approaching clouds of War caused considerable problems for the Vlasov Group, which had vessels registered in opposing countries. Alexandre Vlasov, in a desperate attempt to save his fleet, gave further proof of his skill and brilliant mind, by transferring some of the ships to the flag-of-convenience registry of Panama under the nominal ownership of the Dolphin S.S. Co. Now a common practice among shipowners for economic reasons, in 1941 it was something very unusual.

Fleeing from a Europe in the grip of conflict, the Vlasov family moved to New

York in late 1940 and later to Buenos Aires. As had always happened in his life, Alexandre founded new companies in the countries where he lived: in the United States he opened the Alvion S.S. Corp. Inc. while in Argentina he founded the Compañía Argentína de Navegación de Ultramar (Canumar S.A.) and a small subsidiary operating local fishing trawlers, Compañia Sud Americana de Export & Import S.r.l. (Cosadex).

By the end of the hostilities in 1945 only six vessels had survived; *Tropicus*, *Omega*, *Mimosa*, *Nadin*, *Starstone* and the *Lodestone*. Once again showing his cleverness Mr Vlasov was one of the few private shipowners to obtain a large number of standard war-built freighters: four Victories, three Liberties, a C3 aircraft-carrier from the American Commission, and an Empire vessel from the United Kingdom.

These ships started the post-War freight service of the Group, initially tramping or later on charter, as a result of the enormous demand for freight capacity caused by the devastation of War.

1956 marked the start of the post-War fleet renewal programme, with four new "Stones" being built in Germany and Italy. Years later, in the second half of the 'seventies, three of them would be transferred to Amar Line, a subsidiary of Arabian Maritime S.S. Co., for further trading in Arabian waters. Amar Line had been founded as a 50% joint venture between Vlasov and the Arabian entrepreneur Gaith Pharaon. Amar was the first Western company to be headquartered in Jeddah.

In the late 'sixties the Vlasov Group took over the Italian government contract for the Savona-Vancouver fast freight route, previously controlled by the Italian company Italnavi. For this purpose in 1970 a brand-new company was constituted, Ital Pacific Line, and four modern general cargo freighters of 14,000 dwt ordered from Howaldtswerke AG of Hamburg. Unfortunately the life of this new company was all too brief. In less than two years, owing to the rapid expansion of containerisation, Ital Pacific was closed down. Although they were outdated when compared with the first generation container ships, the four freighters were probably the most modern and efficient type of general dry cargo vessels ever built.

It seems worth remembering that the design was supervised by Boris Vlasov in person and that these ships sported a very high standard of automatisation. He had, for instance, patented a joy-stick control which permitted a single man to operate the cargo derricks which greatly reducing the time needed to handle the cargo and the number of crew involved.

FLEET LIST DATA READING KEY

NAME (YEARS OF SERVICE FOR VLASOV GROUP)
Type of vessel (former names, subsequent names)
Builders: name (place of building)
gross tonnage (net tonnage) deadweight;
length overall [length between perpendiculars] x beam moulded x full load draught
number of engines and type; Brake (oil engines), Indicated (steam reciprocating), Shaft (turbines) or
Nominal HP; service speed
name of engine builders (place where built)

Note: as in the history of the Vlasov Group there are many vessels sharing the same name, a chronological number is given after the ship's name to facilitate her identification. This fleet list contains only the vessels owned by the Vlasov Group and which operated for it. An average of five/six vessels per year is presently bought and sold as part of the company's shipbroking activity; these ships, which do not operate under the Vlasov banner, as well as chartered vessels, are not included.

MIMOSA (1933 - 1949)

(formerly *Kirnwood*, *Michael*, *Mimoza*)
Builders: R. Craggs & Sons Ltd (Middlesbrough)
3071 grt (1950 nrt)
{100.62} x 13.33 x 6.77 m ({330.1} x 47.0 x 22.2 ft)
1 3Exp. Steam Engine; 273 SHP; 10 kn
by Blair & Co. (Stockton)

1905 July: delivered to Constantine & Pickering Steamship Co. as *Kirnwood*; p.o.r. Middlesbrough.
1917: transferred to Joseph Constantine ownership.
1921: transferred to Wood Line Ltd.
1922: sold to the Greek shipowner Demitrios Logothetis and re-named *Michael*; p.o.r. Andros.
1933 February: jointly bought by A. Vlasov and P. Argyropoulos, re-named *Mimoza* and registered in Syra; C. Arvanitides, Piraeus, manager. February 28th: sailed from Piraeus on her first voyage for Vlasov; mainly used in Mediterranean waters for the transport of coal. November: transferred to Vlasov's newly established Scomar, Soc. Commerciale et d'Armement S.A.; re-named *Mimosa*.
1940 November 26th: she left Gibraltar and the Mediterranean waters fleeing from the War for a round voyage to Baltimore. Upon her return she spent many months in the Cape Verde Islands, Lisbon and British ports.

1941 February 22nd: she was caught in a severe Atlantic cyclone, lasting one week; bridge and wheelhouse were carried away, while part of the deck cargo pressing upon the steering rods had to be jettisoned to prevent her going ashore; the vessel managed to arrive at Lisbon where she was repaired.
May 10th: heavily damaged in air raids while in Liverpool: at first the vessel seemed unrecoverable but she was eventually repaired, clearing Liverpool on the following 14th August.
1945 December 15th: after spending the rest of the War in British coast-wise service she entered Piraeus; after calling at Genoa and Barcelona, she crossed the Atlantic bound for New York and Buenos Aires.
1946 August 16th: after repairs and general overhaul she was transferred to Canumar under the Panamanian flag and re-entered service from Buenos Aires, working along the East coast of South America.
1949 March 23rd: she put back to Recife, which she cleared the day before, with serious troubles to her 45 year old reciprocating plant.
On 9th June, she left Recife and slowly crossed the Atlantic arriving on 24th July in Genoa where she was laid up and later sold to Giuseppe Ricordi for scrap.
1950 May 22nd: arrived at Vado Ligure (Savona) to be broken up in the Mario Bertorello scrapyard.

*The Mimosa during her early period of service. Note the pre-War funnel marking of Scomar.
(William Schell, Holbrook)*

FREIGHTERS AND BULK CARRIERS

NADIN (1933 - 1950)

(formerly *Eastwood, Dragon, Nadine*)
Builders: William Gray & Co.
(West Hartlepool) 3582 grt
(2335 nrt) 6030 dwt
[103.63] x 14.63 x 7.56 m
([340.0] x 48.0 x 24.8 ft)
1 3Exp. Steam Engine; 296 SHP; 10 kn
by Central Marine Eng. Works
(West Hartlepool)

1904 March 24th: keel laid. August 29th: launched as *Eastwood*. October 4th: delivered to Eastwood Steamship Co., a subsidiary of Macbeth, Blackwood & Laurie; p.o.r. Hartlepool.
1927: sold to J. D. Corcodilos & Sons Coal Trading & Shipping Co., Piraeus and re-named *Dragon*; Greek flag.
1933: sold to Vlasov's newly established Scomar, Soc. Commerciale et d'Armement S.A.; re-named *Nadine*, then *Nadin* during the same year, after Alexandre Vlasov's stepdaughter who eventually became Boris' wife; p.o.r. Piraeus; fitted for oil fuel.
1934 April 4th: entered service for Scomar, clearing Piraeus for Naples and Istanbul. Used for the transport of coal in Mediterranean waters.
1940 January 8th: after the war broke out she sailed from Istanbul for a long trans-Atlantic trip to South America, returning to Gibraltar via New York on 19th January 1941.
1941 May 24th: seriously damaged by fire with her fleet-mate *Mimosa* during an air raid on Liverpool docks.
1942 May 20th: after repairs resumed service completing a round voyage to Quebec; later used in coastwise service between British ports.
1944 March 27th: commissioned by the Ministry of War Transport, she set sail from Blythe for the first of two voyages to Reykjavik; released from H.M.S. on 31st July. August 22nd: lost her bow in collision with another vessel but reached Belfast where was repaired.
1947 August 14th: arrived at Buenos Aires and transferred to Canumar under the Panamanian flag; used in American waters exception made for a long trans-Atlantic crossing to Genoa with many intermediate calls in winter 1947-1948.
1949 April 6th: sailed Buenos Aires for her last commercial trip; upon her arrival at Genoa, on 31st May, she was laid up.
1950 May 9th: arrived in tow at Savona from Genoa to be broken up.

BORIS II (1933 - 1943)

Standard A type freighter of the War Class
(formerly *War Heather, Menapier*)
Builders: D. & W. Henderson & Co. (Glasgow)
5166 grt (3071 nrt)
[122.05] x 15.94 x 8.69 m
([400.4] x 52.3 x 28.5 ft)
1 3Exp. Steam Engine; 490 SHP; 11 kn by builders

1917 December: delivered to the British Ministry of War Transport, one of the nine War Class standard freighters built by D. & W. Henderson & Co.
1919: sold to Lloyd Royal Belge S.A. and re-named *Menapier*; fitted with tanks for transport of oil; p.o.r. Antwerp.
1933: sold to Vlasov's newly established Scomar, Soc. Commerciale et d'Armement S.A.; re-named *Boris*; p.o.r. Piraeus.
1938 November 24th: while in Hamburg port she collided with another freighter causing the hull to leak; part of the cargo had to be jettisoned to prevent sinking; later repaired.
1939 January 18th: set sail from Istanbul and left the Mediterranean waters; employed on trans-Atlantic voyages between British and American ports.
1940 September 7th: detained by naval authorities

The *Nadin* in Buenos Aires in a 1940 photo.
(William Schell, Holbrook)

at Victoria, B.C. until 26th November.
1943 June 3rd: torpedoed and sunk by a German U-Boat off Ascension Island while en route from Santos to Belfast.

OLTUL - ESMERALDA (1934 - 1943)

(formerly *Dansborg*, then *Rio Deseado*, *Japery*)
Builders: Richardson, Duck & Co. Ltd (Stockton) 4308 grt (2599 nrt)
[115.88] x 15.54 x 7.56 m
([380.2] x 51.0 x 24.8 ft)
1 3Exp. Steam Engine; 397 SHP; 10 kn
by Blair & Co. Ltd (Stockton)

1920 July 15th: launched; in October delivered as *Dansborg* to A/S D/S Dannebrog, p.o.r. Copenhagen, C.K. Hansen managers.
1933 October: sold to Romania Prima Societate de Navigatione Nationala, Bucharest and re-named *Oltul*, p.o.r. Braila. November 11th: left New York, where was laid up, for Braila.
1934 May 30th: laid up at Braila. September 18th: transferred to Alexandre Vlasov Societate de Navigatione, Bucharest and resumed service as coal transport from Polish and German ports to Italian, East Mediterranean and Black Sea ports.
1936 March 8th: collided and run aground in the Kiel Canal; repaired in Gdynia. In June despatched to South America for local trade; on 13th November ran aground in River Plate; refloated on the following 18th.
1937 July: returned to Mediterranean waters via New York.
1938: two voyages to Ceylon via Suez Canal and one transatlantic crossing to New York; in December returned to South America.
1939 December 2nd: cleared Miami bound for Genoa; the following February sailed for Balboa, C.Z. and Recife.
1940 August 9th: arrived at Pernambuco with tank leaks and hull damage caused by a storm; while undergoing repairs Romania entered the war; laid up.
1941 February: transferred to Vlasov's newly established Dolphin Shipping Corp. Panama. September 20th: sailed for Buenos Aires. November 21st: transferred to Vlasov's newly established Compañía Argentína de Navegácion de Ultramar, headquartered in Buenos Aires, becoming its first vessel; re-named *Esmeralda*, p.o.r. Buenos Aires.
1943 October: sold to the Argentine Flota Mercante de Estado, re-named *Rio Deseado*.
1950: sold to Companhia Commercio et Navegaçao, Rio de Janeiro; re-named *Japery*.
1957: sold to Navegaçao Mercantil S.A.; same name.
1968: broken up in Rio de Janeiro.

SIRETUL (1934 - 1953)

(formerly *Baharistan*, then *Omega*, *Castelbruno*, *Castel Bruno*)
see *Castelbruno*, page 44

PRAHOVA (1934 - 1953)

(then *Tropicus*, *Cloverbrook*, *Tropicus*, *Castelverde*, *Castelmarino*)
see *Castelmarino*, page 49

CAMPDEN HILL (1936 - 1937)

Standard freighter of the British War class
(formerly *War Earl*, *Rosyth Castle*, *Umlazi*, then *Okuju* Maru, *Okuzyu* Maru, *Hokuju Maru*)
Builders: Canadian Vickers Ltd (Montreal)
4328 grt (2580 nrt) 7200 dwt
[115.95] x 14.99 x 8.10 m
([380.4] x 49.2 x 26.6 ft)
1 3Exp Steam Engine; 474 NHP; 11 kn
by builders

1918 June 18th: launched as *War Earl*, one of the six standard War class vessels of the 'A' type built by Canadian Vickers to the order of the British Ministry of War Transport. August: completed and delivered.
1919: sold to Union-Castle Line and re-named *Rosyth Castle*, together with another two War class freighters, the *War Soldier* (re-named *Ripley Castle*) and the *War Climax* (re-named *Banbury Castle*).
1920: ownership transferred to Bullard, King & Co. Ltd (Natal Line) and re-named *Umlazi*.
1936 September: sold to the newly established

The *Rio Deseado*, former *Oltul* and (above) the same vessel on her trials as *Dansborg* in October 1920.
(Alberto Bisagno, Genoa)

FREIGHTERS AND BULK CARRIERS

In 1936 the *Umlazi* was bought by Vlasov to became his first British-registered vessel, the *Campden Hill*.
(Laurence Dunn, Gravesend)

Campden Hill Steam Ship Co. Ltd, London as *Campden Hill*; this company was jointly owned by Alexandre Vlasov and the well-known Greek shipowner Manuel Kulukundis (his family still owns a large cargo vessel fleet); the vessel's management was entrusted to Counties Ship Management Co. Ltd (Bury Court House, London) whose ships had all their names ending in Hill.
October 1st: first voyage for her new owners; she cleared Middlesbrough for Gdynia where she loaded coal bound for Costanza, Black Sea; used on the coal route between North Europe and Mediterranean ports.
1937 November 3rd: sold to Kitagawa Sangyo K.K., Tokyo and re-named *Hokuju Maru*, after the closing down of Campden Hill S.S. Co. Ltd. December 1st: sailed Liverpool flying the Japanese flag for Tokyo, where she arrived on the following 2nd April. She started a busy career as tramp vessel (her name being re-spelled *Hokuzyu Maru* in 1938) and was one of the few Japanese ships to survive the Second World War, being used during the conflict as a troop transport between Imari, Japan and Shanghai.
1950: name reverted to *Hokuju Maru*; same owners.
1953: sold to Osaka Zosensho KK; same name; p.o.r. Osaka.
1954: she was in the news for having been converted into a salvage vessel to recover some of the 125 Japanese ships sunk off the Vietnam coast during WW2. In February she sailed from Osaka for Saigon with 180 salvage workers, their boats and 40 lighters. Upon her arrival in Saigon the Vietnamese authorities, however, denied permission to start the salvage works; as a result her owners went bankrupt. November: sold to Kyokuyo Whaling Co. and transformed into a tender vessel for their whaling ships.
1961: her owners abandoned the whaling activity, becoming Kyokuyo Ltd and the vessel was laid up in Osaka.
1964: broken up in Osaka.

SUNSTONE I (1936 - 1942)

(formerly *Inverleith*, then *Castelverde*)
see *Castelverde*, page 53

PEARLSTONE I (1936 - 1941)

(formerly *Zapala*, *Ovingdean Grange*, then *Castelnuovo*, *Castelbianco*)
see *Castelbianco*, page 57

GEMSTONE I (1938 - 1942)

Builders: James Laing & Sons Ltd (Sunderland)
4986 grt (2941 nrt); 125.81 x 17.60 x 7.32 m
(412.8 x 57.7 x 24.0 ft)
1 3Exp Steam Engine; 550 NHP; 12 kn
by North Eastern Marine Eng. Co. Ltd (Sunderland)

1937: Ordered by Minster Steamship Co. Ltd, London with a sistership; bought by A. Vlasov while on the stocks. Launched on 30th December as *Gemstone* for Alva S.S. Co. Ltd, London.
1938 February 7th: Delivery trials; started service under Navigation & Coal Trade Ltd management.
1942 June 4th: Sunk by the German raider *Stier* at 1°52"N, 26°48"W while en route from Durban to Baltimore with a load of manganese ore; 24 of her 37 crew members were saved and imprisoned.

After her sale to Japan, the *Campden Hill* became the *Hokuju Maru*. Here she his seen during her WWII duty as troop transport.
(Sanae Yamada, Tokyo)

Freighters and bulk carriers

The *Starstone* as an auxiliary British patrol vessel during the Second World War.
(National Maritime Museum, Greenwich)

The 1938-built *Starstone* was the first newbuilding ordered by the Vlasov Group.
(Alex Duncan, Gravesend)

A post-War aerial view off Dover of the first *Lodestone*.

FREIGHTERS AND BULK CARRIERS

STARSTONE I (1938 - 1963)
Builders: William Doxford & Sons Ltd (Sunderland)
5702 grt (3441 nrt); 131.60 x 17.50 x 8.10 m
(431.8 x 57.4 x 26.6 ft)
1 3Exp Steam Engine; 550 NHP; 12 kn
by Richardson, Westgarth & Co. Ltd (West Hartlepool)

1938 May 14th: Launched. Delivered to Alva S.S. Co. Ltd, London and entered service under Navigation & Coal Trade Ltd management.
1959: transferred to the Alvada Shipping Co. Ltd, Hamilton.
1963 February 4th: arrived at Nagoja, Japan to be broken up.

LODESTONE I (1938 - 1963)

Builders: Bartram & Sons Ltd (Sunderland)
4877 grt (2887 nrt); 126.61 x 16.92 x 7.64 m
(415.4 x 55.5 x 25.1 ft)
1 3Exp Steam Engine; 550 NSHP; 12 kn
by North Eastern Marine Eng. Co. Ltd (Sunderland)

1938 July 27th: Launched. Delivered to Alva S.S. Co. Ltd, London and entered service under Navigation & Coal Trade Ltd management.
1959: transferred to the Alvada Shipping Co. Ltd, Hamilton.
1963: broken up in Japan.

ALMERIA (1940 – 1940)

(formerly *Lake Slavi, Almeria Lykes*)
Builders: Great Lakes Engi. Works (Ecorse)
2637 grt (1762 nrt); 74.19 x 13.35 x 7.92 m
(243.4 x 43.8 x 26.0 ft)
3 Exp. steam engine; 350 NHP; 11 kn
by builders (Detroit)

1920 January: delivered as *Lake Slavi* to the U.S. Government for the freight service on Lake Michigan; standard "Lake" class vessel.
1922: sold to Lykes Brothers and re-named *Almeria Lykes*; p.o.r. Galveston.
1936 March: refitted; nrt increased from 1658 to 1762 t.
1938 December 23rd: laid up in Galveston after suffering collision damage in the Houston Channel.
1940 March 29th: sold to the Sitmar Line; name altered to *Almeria*. May 5th: sailed from Galveston for Montevideo. May 19th: wrecked off the north coast of Trinidad before her first voyage under the Sitmar banner was completed; total loss.

POTENZA (1943 - 1944)

(formerly *Auvergne*)
Builders: J. Samuel White & Co. Ltd (Cowes)
2114 grt (1203 nrt) 3275 dwt
[85.34] x 12.43 x 5.88 m
(280.0 x 40.2 x 19.3 ft)
1 3Exp. steam engine; 243 NHP; 10 kn
by builders

1921 October: completed and delivered as *Auvergne* to Compagnie Delmas Vieljeux; p.o.r. La Rochelle.
1942 November: requisitioned by the Italian Navy. December 21st: delivered to the Italian Line and re-named *Potenza*; p.o.r. Genoa.
1943 April 5th: transferred to the Sitmar Line, in replacement of the *Castelverde*, lost the previous December.
1944 August 20th: sunk at Port-Saint-Louis-Du-Rhône, Marseilles during War action.

CALTANISETTA (1942 - 1943)

(formerly *Lilian Spliethoff, La Tamise, Impöna, Tamise*)
Builders: U. Zvolsman & Zoon (Workum)
265 grt (146 nrt); 36.27 x 6.99 x 3.54 m
(119.0 x 23.0 x 11.6 ft)
1 2Exp. steam engine; 16 NHP; 8 kn
by Botje Ensing & Co. (Groningen)

1917: built and delivered as *Lilian Spliethoff* to the Dutch shipowner Spliethoff.
1919: sold to the French company Paris-Londres Maritime S.A. and re-named *La Tamise*.
1924: sold to Compagnie Côtiere de Madagascar and re-named *Impöna*.
1927: sold to Societé des Ancien Etablisment Courbet Frères, Marseilles, and re-named *Tamise*.
1938: sold to Spada & Lassale, Nice; same name.
1942 November: seized by Italy in Nice. December 14th: delivered to the Italian Line and re-named *Caltanisetta*, p.o.r. Genoa.
1943 May 19th: transferred to Sitmar Line while in Taranto.
1944 May: sunk in Viareggio during War action.

The *Lake Slavi*, became Vlasov's *Almeria* in 1940, but she had a very short stint under the new banner, being wrecked on her first voyage for Sitmar Line. *(Peter Newall, Blandford Forum)*

BESUGO (1943 - 1962)

Fishing trawler
(formerly *Graf Bothmer, Star of Liberty, Graf Bothmer, Marie Sprenger*)
Builders: Reiherstieg Schiffwerk (Hamburg)
215 grt (96 nrt); [36.58] x 7.41 x 2.93 m
([120.0] x 24.3 x 9.6 ft)
1 3Exp Steam Engine; 40 NHP; 8,5 kn
by builders

1918: keel laid to the order of Kaiserlische Marine as yard no. 502.
1919 January: completed and laid up.
1920: taken over by German Government; in May delivered to the British Government as War reparation and transferred to Hull.
1922: sold to Walker Steam Trawler Fishing Co. Ltd, Aberdeen, together with the similar vessel *Reiherstieg*; re-named *Star of Liberty* and fitted as fishing trawler.
1923 June: sold to Sirius Handels-Gesellschaft, Bremerhaven; name reverted to the original *Graf Bothmer*; later that year re-sold to Von der Heide, Altona and re-named *Marie Sprenger*.
1928: sold to Industria Pesquadora Argentina, Buenos Aires; re-named *Besugo* and used as fishing trawler in Argentine waters.
1939: sold to Società Pesquadora Cardella, Buenos Aires, same name.
1943 September: sold to Vlasov's newly established Cosadex (Compañía Sud Americana De Export & Import, S.R.L.) together with the similar vessel *Lenguado*.
1947: fitted for oil fuel.
1962 August 28th: sold for scrap together with the *Lenguado*.

LENGUADO (1943 - 1962)

Fishing trawler (formerly *Reiherstieg, Star of the Realm, Rheierstieg, Elisabeth Sprenger*)
Builders: Reiherstieg Schiffwerk. (Hamburg)
236 grt (88 nrt); [36.81] x 6.79 x 4.08 m
([120.8] x 22.3 x13.4 ft)
1 3Exp Steam Engine; 40 NHP; 8 kn
by builders

1915: keel laid as yard no. 490 on builders' account.
1917: completed and commissioned by the Kaiserliche Marine as patrol vessel.
1919: delivered to the British Government as War reparation.
1922: sold to Walker Steam Trawler Fishing Co. Ltd, Aberdeen, together with the similar vessel *Graf Bothmer*; re-named *Star of the Realm* and fitted as fishing trawler.
1923 June: sold to Sirius Handels-Gesellschaft, Bremerhaven; name reverted to the original *Reiherstieg*; later that year re-sold to Von der Heide, Altona and re-named *Elisabeth Sprenger*.
1928: sold to Industria Pesquadora Argentina, Buenos Aires; re-named *Lenguado* and used as fishing trawler in Argentine waters.
1939: sold to Società Pesquadora Cardella, Buenos Aires, same name.
1943 September: sold to Vlasov's newly established Cosadex (Compañía Sud Americana De Export & Import, S.R.L.) together with the similar vessel *Besugo*.
1947: fitted for oil fuel and with a refrigerated hold.
1962 August 28th: sold for scrap together with the *Besugo*.

GIAMPAOLO (1947 – 1949)

(formerly *Hebron, Wm. Minlus II, Aldebaran, Chrissula*, then *Maristella*)
Builders: Schiffs & Dockbauwerft Flender Siems AG (Lübeck)
338 grt (180 nrt) 450 dwt
44.80 [42.39] x 6.85 x 3.80 m
([144.03] x 22.47 x 12.47 ft)
1 2Exp Steam Engine); 225 SHP; 8 kn
by Seebeckwerft AG (Geestemünde)

1920 November 22nd: launched as the *Hebron* to the order of Hamburg America Line, fourth of six sisterships intended for the freight service in the North Sea.
1921 February 17th: delivered; entered service four days later on the route between German and Scandinavian ports; however, one year later, Hapag closed down its freight service in the North Sea and the six new coasters were put up for sale.
1922 March 3rd: sold to the Lübeck shipowner Wilhelm Minlos together with her sister ship *Lebanon*. The latter became the *Wm. Minlos I* while the *Hebron* became the *Wm. Minlos II*, although she remained in lay-up. April 6th: re-sold to Deutsch Skandinavische Reederei AG, Hamburg. April 25th: entered service for her new owner with the name *Aldebaran* on the German-Scandinavian route.
1925 March 12th: sold to the Italian shipowner Evangelista Boumis of Syracuse and re-named *Chryssula*; used in Mediterranean inter-coastal service.
1932: sold to Cav. Onorato Gorlato of Muggia, Trieste; re-christened *Giampaolo* and used in the

The Giampaolo was the first post-War acquisition of Sitmar Line, although she remained in lay-up until re-sold.
(Arnold Kludas, Grünendeich)

North Adriatic inter-coastal service.
1937: sold to Industria Marmifera Impresa di Costruzioni e Trasporti, Trieste; used for the trade of marble and Istrian stone along the Adriatic coasts.
1943 September: after Italy's capitulation the *Giampaolo* looked for shelter with many other vessels in Southern Italy, reaching Taranto, which was already in the hands of Allied forces. Requisitioned by the Allied authorities and used for the transport of materials from Malta to Italy.
1946 July 6th: released and laid up in Alexandria.
1947 December 16th: after having been sold to Sitmar Line (becoming its first post-war acquisition), she was brought to Brindisi and laid up, never operating on behalf of Vlasov.
1949: sold to Compagnia di Navigazione Marittima Commerciale, Bari and re-named *Maristella*; used for the transport of Calcium Carbonate between Turkey and Italy.
1950 February 16th: while en route from Bagnoli, Naples to Savona she was caught in a storm and sank in shallow waters near Civitavecchia.
1951 May 31st: refloated and towed to the port of Civitavecchia. June 24th: re-launched after having been re-hauled on a slipway of a local shipyard and repaired.
1954 January 31st: disappeared with her crew and a load of Calcium Carbonate while en route from Erchie, Turkey, to Bagnoli, Italy.

RUBYSTONE I (1947 - 1960)

(formerly *Monte Piana*, *Empire Baron*)
Builders: Cantiere Navale Triestino (Monfalcone)
5894 grt (3718 nrt); [121.92] x 16.15 x 8.45 m
({400.0] x 53.0 x 27.7 ft)
1 single-acting 6-cyl. Diesel engine; 1950 SHP; 10.5 kn
by Stabilimento Tecnico Triestino (Trieste)

1925 February 9th: laid down as Yard no. 156 to the order of Gerolimich & Co., Trieste. Sistership to the *Col di Lana* (Yard. no. 155, same builders); launched on the following 20th August as the *Monte Piana*.
1926 June 8th: delivered.
1932 August 10th: first voyage Genoa-Las Palmas-Rio de Janeiro-Santos-Montevideo in charter to the Italian Line, as a replacement to the *Caprera*, which went aground 15 miles off Rio de Janeiro on the previous 1st June.
1934 January 20th: delivered back to her owner after the completion of six round voyages to South America for the Italian Line.
1940 June 10th: when Italy entered the War the vessel was at Aden; her crew wrecked the oil engine and tried to scuttle the vessel but on the 12th she was eventually beached by the Royal Navy before she sank.
1941 February 23rd: arrived in tow by the steamer *Nirvana* at Vizagapatam, India; repaired and put in service for the British Ministry of War Transport with the name *Empire Baron*.
1947: sold to Alva S.S. Navigation Co. Ltd, London and re-named *Rubystone*, p.o.r. London.
1951 January: transferred to the Alvion Steamship Corp., Panamanian flag.
1960 August 27th: arrived at Nagasaki, Japan to be broken up.

CORALSTONE - ESMERALDA II (1947 - 1968)

Liberty Ship
(formerly *Priscilla Alden*, *Samlouis*)
Builders: Bethlehem Fairfield S.B. Inc. (Baltimore)
7261 grt (5230 nrt) 10981 dwt; 134.57 [126.80] x 17.37 x 10.56 m
(441.5 {416.0] x 57.0 x 34.6 ft)
1 3Exp. Steam Engine; 2500 IHP; 11 kn
by General Machinery Corp. (Hamilton)

The *Coralstone* was one of the three Liberty Ships obtained by Vlasov from the U.S. Maritime Commission to re-constitute his post-War fleet of freighters.
(National Maritime Museum, Greenwich)

In 1959 the Liberty Ship *Coralstone* was transferred from Alva S.S. Co. Ltd of London to Canumar of Buenos Aires, becoming the *Esmeralda*. (Alberto Bisagno, Genoa)

1943 October 15th: keel laid as yard no. 2261. October 25th: launched as *Priscilla Alden*. November 13th: sailed Baltimore entering service with the name changed to *Samlouis* for British Ministry of War Transport, Ellerman's Wilson Line Ltd appointed managers, p.o.r. London. Used in Indian, Mediterranean and South American waters.
1947 May 2nd: bought by Alva and re-named *Coralstone*, p.o.r. London, Navigation & Coal Trade Ltd managers. May 9th: sailed London for Montreal and Bombay in world-wide tramp trade. June 15th: lost propeller off St. Pierre Island, towed to Halifax for repairs.
1951 January 10th: transferred to the Alvion Steamship Corp., Panamanian flag.
1952 July 13th: laid up in Genoa; re-entered service on the following 29th January after refurbishing works. August 27th: fire in her cargo of coal while in Buenos Aires; flooded and towed to yard for repairs.
1959 September 29th: transferred to Canumar ownership, p.o.r. Buenos Aires, and re-named *Esmeralda*; sailed Buenos Aires on the following 21st.
1966 May 8th: ran aground off Montevideo; refloated two days later.
1967 July 24th: laid up at Buenos Aires and put up for sale.
1968 October 3rd: arrived at Kaohsiung from Buenos Aires to be broken up.

IRA (1947 - 1947)

Liberty Ship (formerly *Harry Percy*)
Builders: Todd S.B. Corp. (Houston)
7244 grt (4396rt) 10900 dwt
134.57 (126.80) x 17.37 x 10.56
m (441.5 (416.0) x 57.0 x 34.6 ft)
1 3Exp. Steam Engine; 2500 IHP; 11 kn
by General Machinery Corp. (Hamilton)

1943 November 14th: keel laid. December 21st: launched. December 31st: delivered to U.S. War Shipping Administration as *Harry Percy*; States Marine Corp. managers.
1947 January 7th: sold to Scomar and re-named *Ira*, p.o.r. Piraeus. March 7th: while en route from Hampton Road to Antwerp on her first voyage for Vlasov with a load of coal she ran aground on the Goodwin Sands, off the south east coast of England, very close to the wreck of the *Luray Victory*. Upon request of her master, Capt. Efthymios Zisimos, the tugs *Lady Duncannon* and *Lady Brassey* arrived on the scene in a vain attempt to refloat her. Later the upperworks collapsed and her back was broken. The 34 members of her crew were all saved. Salvaging of the wreck as well of the cargo considered impracticable; total loss.

The ill-fated *Ira*, in a photograph taken by the U.S. Coast Guard on 20th January 1944, when she was the brand new Liberty ship *Harry Percy*. (The Mariners' Museum, Newport News)

Freighters and bulk carriers

The Olga during her last period of service with the traditional V painted on her funnel.
(Alex Duncan, Gravesend)

OLGA (1947 - 1968)

Liberty Ship (formerly *Susan Colby*)
Builders: New England S.B. Corp. (Portland)
7244 grt (4380 nrt) 10900 dwt
134.57 {126.80} x 17.37 x 10.56 m
(441.5 {416.0} x 57.0 x 34.6 ft)
1 3Exp. Steam Engine; 2500 IHP; 11 kn
by Harrisburg Machinery Corp. (Springfield)

1943 November 24th: keel laid.
1944 January 13th: launched. January 26th: delivered to U.S. War Shipping Administration as *Susan Colby*. Moore-McCormack Lines Inc., New York, appointed managers.
1947 January 7th: sold to Scomar and re-named *Olga*, p.o.r. Piraeus, Navigation & Coal Trade Ltd management.
1950: transferred to Alva S. S. Co., London.
1967 October 16th: in a voyage from Alexandria to Bombay she put in to Cape Town with a severe crack in her hull and the subsequent flooding of her holds; laid up.
1968 August: arrived at Kaohsiung to be broken up.

A February 1948 photograph of Olga, sporting the post-War Scomar funnel marking and leaving Cape Town with a full load of timber.
(Len Sawyer & Victor Young, Wellington)

FREIGHTERS AND BULK CARRIERS

The Argentina Victory, one of the four Victories ships bought by Vlasov in January 1947, was re-sold only one year later to Holland America Line to became their Akkrumdijk (below).
(National Maritime Museum, Greenwich and Skyfotos, New Romney)

ARGENTINA VICTORY (1947 - 1948)

Victory Ship (formerly *Lake Charles Victory*, then *Akkrumdijk*, then *Akkrumdyk*, then *Pacific Comet*)
Builders: Bethlehem Fairfield S.B. Corp. (Baltimore)
7639 grt (4571 nrt) 10800 dwt
138.77 {133.84} x 18.93 x 10.52 m
(455.3 {439.1} x 62.1 x 34.5 ft)
1 set of H.P. and L.P. DR geared turbines; 8500 SHP; 17 kn
by Westinghouse E. & M. Co. (Pittsburgh)

1944 December 5th: keel laid.
1945 February 1st: launched. February 28th: delivered and entered service for the U.S. War Shipping Administration as *Lake Charles Victory*.
1947 January: sold to Canumar together with her sister ships *Vassar Victory*, *Wooster Victory* and *Rollins Victory*; re-named *Argentina Victory*.

1948 August 28th: sold to Holland America Line; re-named *Akkrumdijk*. With her former Vlasov-owned sistership *Patagonia Victory* she joined a fleet of another 8 Victories owned by H.A.L.; another two were acquired in 1950 and 1952, making a total number of 12; all their names began with A and terminated with -dijk, changed in 1954 to -dyk.
1949 November: on the Bremen - U.S.A. regular line.
1954: re-named *Akkrumdyk*.
1962 April 30th: sold to Overseas Maritime Inc. Monrovia; re-named *Pacific Comet*, Liberian flag.
1968 November 20th: engine broke down in the Pacific after heavy damages caused by a storm; towed to Honolulu; resumed the voyage to Saigon at reduced speed on the H.P. turbine only; laid up.
1969 May 5th: arrived at Kaohsiung to be broken up.

PATAGONIA VICTORY (1947 - 1948)

Victory Ship (formerly *Rollins Victory*, then *Arendsdijk*, *Arendsdyk*, *Hongkong Exporter*)
Builders: Bethlehem Fairfield S.B. Corp. (Baltimore)
7639 grt (4571 nrt) 10800 dwt
138.77 {133.84} x 18.93 x 10.52 m
(455.3 {439.1} x 62.1 x 34.5 ft)
1 set of H.P. and L.P. DR geared turbines
8500 SHP; 17 kn
by Westinghouse E. & M. Co. (Pittsburgh)

1945 May 17th: keel laid. July 2nd: launched. July 31st: delivered and entered service for the U.S. War Shipping Administration as *Lake Charles Victory*.
1947: sold to Canumar together with her sister ships *Vassar Victory*, *Wooster Victory* and *Lake Charles Victory*; re-named *Patagonia Victory*.
1948 August 17th: sold to Holland America Line;

Freighters and bulk carriers

Montevideo, 2nd November 1947: the *Patagonia Victory* during her short career with Vlasov. *(William Schell, Holbrook)*

re-named *Arendsdijk*. With her former Vlasov-owned sistership *Argentina Victory* she joined a fleet of another 8 Victories owned by H.A.L.; another two were acquired in 1950 and 1952, making a total number of 12; all their names began with A and terminated with -dijk, changed in 1954 to -dyk.
1954: re-named *Arendsdyk*.
1961: sold to International Export Lines Ltd, Nassau; re-named *Hongkong Exporter*.
1972 April 7th: cleared Inchom on her last voyage. July 15th: delivered to the demolition firm of Nan Feng Steel Enterprise Co., Kaohsiung.

CORVINA (1952 - 1961)

Fishing Trawler
(formerly *Wappen von Hamburg*, *Adala*)
Builders: H. Brandenburg (Hamburg)
261 grt (109 nrt); 45.96 x 6.80 x 4.08 m
(150.8 x 22.3 x 13.4 ft)
1 3Exp Steam Engine by builders: 43 NHP; 8 kn by builders

1906 October 17th: launched as *Wappen von Hamburg*. November 21st: completed and delivered to Hochseefischerei AG Hansa, Hamburg.
1914 August 3rd: commissioned by the Kaiserlische Marine as coast patrol boat in the North Sea.
1918 November 24th: released from war service.
1925 May 18th: laid up in Altona and put up for sale.
1926: bought by Canto y Ribattu, Compañía Argentina de Pesca y Derivacos Termao Quimicos S.A., Buenos Aires; re-named *Adala*.
1930: sold to Industria Pesquera Argentina S.A., Buenos Aires, and re-named *Corvina*.
1937: sold to Soc. Pesquerias Gardella S.A. (Pesgar S.A.); same name.
1938: sold to Josè Maria Castagnino, Buenos Aires; same name.
1948: sold to Productora de Aceite de Pescado S.R.L.; J.C. Schenone, Agencia Maritima managers.
1952: sold to Vlasov's Cosadex, Compañia Sud Americana De Export & Import, S.R.L., joining the two other fishing trawlers owned by the Canumar subsidiary, *Besugo* and *Lenguado*.
1961 November 22nd: caught fire and settled on

The *Corvina* was one of three Argentine fishing trawlers owned by Vlasov's Cosadex of Buenos Aires. *(Helsa H. Lleonart, Buenos Aires)*

37

FREIGHTERS AND BULK CARRIERS

the bottom in Buenos Aires harbour. December 13th: hulk sold for scrap o to V. Montes, Buenos Aires.

NEUQUEN (1956 - 1957)

Standard War Class freighter (formerly *War Column, Attività, Valdarno, Rio Neuquen*)
Builders: J. Cughlan & Sons (Vancouver)
5696 grt (3465 nrt)
130.15 {125.97} x 16.55 x 8.29 m
(427.0 {410} x 54.29 x 27.2 ft)
1 3Exp Steam engine; 2000 SHP; 13 kn
by Stabilimento Tecnico Triestino (Trieste)

1919 May 3rd: launched as *War Column* to the order of the British Ministry of War Transport. Fitted with an HP and LP turbine set by Halliday (Spokene); 2000 BHP, 11,5 kn service speed. July: delivered and entered service.
1920: sold to the Italian company Lloyd del Pacifico, a Zino Line subsidiary, and re-named *Attività*; p.o.r. Savona.
1929 June: re-engined with a second hand 3 Exp. reciprocating steam engine built by Stabilimento Tecnico Triestino in 1914.
1931: sold to Corrado S.A. of Genoa and re-named *Valdarno*; p.o.r. Genoa.
1935 January: chartered until October to the Tirrenia Line to be used as war transport during the Italian campaign in Ethiopia.
1940 June 10th: when Italy entered the WWII the steamer was in Buenos Aires where she was laid up.
1941 August: sold to the Argentine Flota Mercante de Estado; re-named *Rio Neuquen*; p.o.r. Buenos Aires.
1956 July 15th: sold to Canumar and re-named *Neuquen*.
1957 April 4th: cleared Buenos Aires for her last voyage bound for Recife, Dakar, Las Palmas, Algiers and Genoa, where she was laid up on the following 8th June. December 7th: delivered to the demolition yard Cantieri Navali Santa Maria S.p.A., Portovenere (La Spezia).

The second Pearlstone owned by Vlasov was built in 1953 but was sold just one year later; she was the first post-War newbuilding of the Group.

PEARLSTONE II (1953 - 1954)

Bulk carrier
(then *Leto, Costaflora, Pearl Delta*)
Builders: Short Bros. Ltd (Sunderland)
5925 grt (3281 nrt) 5173 dwt
140.54 {136.64} x 18.23 x 7.86 m
(461.1 {448.3} x 59.8 x 25.8 ft)
1 single-acting 6-cyl. Diesel engine
6000 SHP; 14.5 kn
by Harland & Wolff Ltd (Glasgow)

1953 March 17th: launched as *Pearlstone* for Alva S.S. Co. Ltd, London. August: delivered; Navigation & Coal Trade Ltd appointed managers; British flag.
1954: sold to the Dutch company N.V. Maats. Zeevaart, Rotterdam; re-named *Leto*; Hudig & Veder appointed managers.
1962 January: laid up in Rotterdam until May of the following year.
1967: sold to Cia. Mar. Leduma S.A., Piraeus and re-named *Costaflora*.
1979: sold to Pearl Gulf Transportation, Panama and re-named *Pearl Delta*; Thunderbird Pte Ltd, Singapore, appointed managers.
1981 December 9th: arrived Gadani Beach to be broken up.

PEARLSTONE III - STARSTONE II AL BANDARI (1956 - 1983)

General dry cargo vessel
Builders: Nordseewerke AG (Emden)
6213 grt (3627 nrt) 4680 dwt
153.80 {144.63} x 19.23 x 9.14 m
(504.6 {474.5} x 63.10 x 30.0 ft)
1 single-acting 6-cyl. Diesel engine; 6340 SHP; 15 kn
by M.A.N. Maschinenfabrik AG (Augsburg)

1956 August 30th: launched with the name *Pearlstone* for Alvion S.S. Corp., Panama.
1963: transferred to Monrovia Shipping Corp.

FREIGHTERS AND BULK CARRIERS

1965: transferred to the Pearlstone Shipping Corp., Monrovia.
1967: transferred to the Starstone Shipping Corp., Monrovia and re-named *Starstone*.
1981: transferred to Amar Line, Jeddah and re-named *Al Bandari*, Univan Shipping Management, Hong Kong, appointed managers.
1983 October 5th: arrived at Gadani Beach, Pakistan to be broken up.

GEMSTONE II - AL HIJAZI (1957 - 1983)

General dry cargo vessel (then *Al Johffa*)
Builders: Nordseewerke AG (Emden)
6213 grt (3627 nrt) 4680 dwt
153.80 [144.63] x 19.23 x 9.14 m
(504.6 [474.5] x 63.10 x 30.0 ft)
1 single-acting 6-cyl. Diesel engine; 6340 SHP; 15 kn
by M.A.N. Maschinenfabrik AG (Augsburg)

1957 January 29th: launched as *Gemstone* for Alvion S.S. Corp. Panama.
1965 December: transferred to Gemstone Shipping Corp., Monrovia; Liberian flag.
1976: transferred to Amar Line, Jeddah and re-named *Al Hijazi*.
1983: sold to Ahmed Abdul Qawi Bamaodah, Saudi Arabia and re-named *Al Johffa*.
1985 April: arrived at Alang, India to be broken up.

GOLDSTONE (1958 - 1979)

General dry cargo vessel (formerly *Moonstone*, then *Jhanda*, *Bengol Tower*)
Builders: Cantieri Navali di Taranto S.p.A. (Taranto)
8615 grt (5135 nrt) 13457 dwt
148.62 [137.17] x 18.90 x 11.97 m
(487.6 [450.0] x 62.0 x 39.3 ft)
1 single-acting 7-cyl. Diesel engine
6200 SHP; 14.00 kn
by Fiat Grandi Motori (Turin)

1958: launched as *Moonstone* but delivered as *Goldstone* to Alvion Shipping Corp. Panama.
1966 January: transferred to the Goldstone Shipping Corp., Monrovia; Liberian flag.
1979: sold to High Water Navigation Co. Ltd, Dacca (Bangladesh) and re-named *Jhanda*.
1980: sold to Bengol Liner Ltd, Chittagong (Bangladesh) and re-named *Bengol Tower*.
1985 May 5th: laid up in Chittagong. June 15th: demolition started in Chittagong by Chowdhury & Co.

SILVERSTONE - ALASSIRI (1958 - 1983)

General dry cargo vessel (formerly *Sunstone*)
Builders: Cantieri Navali di Taranto S.p.A. (Taranto)
8615 grt (5135 nrt) 13457 dwt
148.65 [137.17] x 18.90 x 11.97 m
(487.7 [450.0] x 62.0 x 39.3 ft)
1 single-acting 7-cyl. Diesel engine
6200 SHP; 14.00 kn
by Fiat Grandi Motori (Turin)

The 1956-built *Starstone* in Genoa.
(Antonio Scrimali, Alpignano)

The second *Gemstone* in Venice.
(Matteo Parodi, Monaco)

FREIGHTERS AND BULK CARRIERS

Among the traditional names of Vlasov's freighters ending in "stone" there was only one *Goldstone*, although she was launched with another name, *Moonstone*, again no more used by Vlasov freighters.
(Antonio Scrimali, Alpignano)

1958: launched with the name *Sunstone* but delivered as the *Silverstone*.
1966 January: transferred to the Silverstone Shipping Corp., Monrovia
1976: transferred to the Amar Line, re-named *Alassiri*, p.o.r. Jeddah; M.C. Shipping Management S.A.M., Monte Carlo.
1983 May 21st: left Jeddah bound for Alang to be broken up by Quality Steel PVT.

RUBYSTONE II (1970 - 1972)

General dry cargo vessel (then *Maracaibo*)
Builders: Howaldtswerke AG (Hamburg)
8335 grt (4797 nrt) 11145 dwt
165.00 [155.00] x 23.60 x 9.82 m (541.3 [508.5] x 77.4 x 32.2 ft)
1 single-acting 8-cyl. Diesel engine; 22000 BHP; 22,5 kn
by Fiat Grandi Motori (Turin)

1969 April: ordered with three sistership (*Coralstone II*, *Lodestone II* and *Pearlstone IV*) to run the Savona-Vancouver express freight service. November 3rd: launched with the name *Rubystone*.
1970 June 17th: maiden voyage for Rubystone Shipping Corp. Monrovia in charter to Ital Pacific Line for their Savona-Vancouver express freight service.
1972: sold to Compañía Anonima Venezolano de Navegaçion, Maracaibo and re-named *Maracaibo*.
1984 June 11th: laid up at Mobile.
1988 April 2nd: arrived in tow at Kaoshiung from Mobile to be broken up.

LODESTONE II (1970 - 1972)

General dry cargo vessel
(then *Amersfoort*, *Lloyd Auckland*, *Nedlloyd Amersfoort*, *Mercury Sky*, *Taal Lake*)
Builders: Howaldtswerke AG (Hamburg)
10239 grt (5818 nrt) 17011 dwt
191.32 [186.70] x 23.60 x 9.82 m
(627.7 [612.5] x 77.4 x 32.2 ft)
1 single-acting 8-cyl. Diesel engine
22000 BHP; 22 kn
by Fiat Grandi Motori (Turin)

1969 April: ordered with three sistership (*Coralstone II*, *Rubystone II* and *Pearlstone IV*) to run the Savona-Vancouver express freight service, after the Vlasov Group took on the Italian government contract previously held by Italnavi.
1970 January 5th: launched as *Lodestone* for Lodestone Shipping Corp., Monrovia and chartered to Ital Pacific Line for their Savona-Vancouver express freight service. She and the *Pearlstone* were longer than their sisterships *Coralstone* and *Rubystone* by the insertion of a 31.70 m (104.0 ft) cylindrical section.
1972: sold together with her sistership *Pearlstone* to Koninklijke Nederlandsche Stoomboot Maatschappij BV, Amsterdam and re-named *Amersfoort*.
1981: re-named *Lloyd Auckland*, same owner.
1982: transferred to Nedlloyd Rederijdiensten BV, Amsterdam and re-named *Nedlloyd Amersfoort*.
1989: sold to Ace River Co. Inc. and re-named *Mercury Sky*; p.o.r. Port Vila, Vanuatu Islands.
1991: sold to A.P. Madrigal Steamship Co. Inc. and re-named *Taal Lake*; p.o.r. Port Vila.
1992 July 4th: demolition started in Chittagong, Bangladesh, by Rashed Enterprise.

FREIGHTERS AND BULK CARRIERS

CORALSTONE II (1970 - 1972)

General dry cargo vessel
(then *La Guaira*, *Bella*)
Builders: Howaldtswerke AG (Hamburg)
8490 grt (4762 nrt) 12340 dwt
165.00 [155.00] x 23.60 x 9.82 m
(541.3 [508.5] x 77.4 x 32.2 ft)
1 s. a. 8-cyl. Diesel engine; 22000 BHP; 22,5 kn
by Fiat Grandi Motori (Turin)

1969 April: ordered with three sisterships (*Rubystone II*, *Lodestone II* and *Pearlstone IV*) to run the Savona-Vancouver express freight service, after the Vlasov Group took on the Italian government contract previously held by Italnavi.
1970 April 15th: launched as *Coralstone* for Coralstone Shipping Corp. Monrovia to be chartered to Ital Pacific Line for their Savona-Vancouver express freight service.
1972: sold to Compañía Anonima Venezolano de Navegaçion, La Guaira and re-named *La Guaira*.
1985 May 1st: while en-route from Maracaibo to Genoa a fire broke out on board killing a crew member; two days later put in at Funchal for the most urgent repairs and then despatched to a Cadiz shipyard for permanent repairs; left Cadiz on the following 4th July.
1990: sold to Bella Shipping Inc. Kingston and re-named *Bella*; St. Vincent & Grenadines flag.
1992 March 21st: demolition started at Port Alang, India by Kalthia & Co. R.L.

PEARLSTONE IV (1971 - 1972)

General dry cargo vessel
(then *Alkmaar*, *Lloyd Auckland*, *Alkmaar*, *Nedlloyd Alkmaar*, *Mercury Sea*)
Builders: Howaldtswerke AG (Hamburg)
10239 grt (5818 nrt) 17011 dwt; 191.32 [186.70] x 23.60 x 9.82 m (627.7 [612.5] x 77.4 x 32.2 ft)
1 single-acting 8-cyl. Diesel engine; 22000 BHP; 22 kn
by Fiat Grandi Motori (Turin)

1969 April: ordered with three sisterships (*Coralstone II*, *Lodestone II* and *Rubystone II*) to run the Savona-Vancouver express freight service, after the Vlasov Group took on the Italian government contract previously held by Italnavi.
1970 July 15th: launched as *Pearlstone*.
1971: completed and delivered to Pearlstone Shipping Corp. Monrovia.
Chartered to Ital Pacific Line for their Savona-Vancouver express freight service.
She and the *Lodestone* were longer than their sisterships *Coralstone* and *Rubystone* by the insertion of a 31.70 m (104.0 ft) cylindrical section.
1972: sold together with her sistership *Lodestone* to Koninklijke Nederlandsche Stoomboot Maatschappij BV, Amsterdam and re-named *Alkmar*.
1980: re-named *Lloyd Auckland*, same owner.
1981: name reverted to *Alkmaar*, same owner.
1982: sold to Nedlloyd Rederijdiensten BV, Amsterdam and re-named *Nedlloyd Alkmaar*.
1989: sold to Ace River Co. Inc. and re-named *Mercury Sea*; p.o.r. Port Vila, Vanuatu Islands.
1993 April 18th: arrived at Chittagong, Bangladesh to be broken up by M.R. Enterprises.

PROGRESO ARGENTINO
LUCAYAN TRADER (1970 - 1986)

Bulkcarrier
Builders: Astilleros Argentinos Rio de la Plata S.A. (Buenos Aires)
9927 grt (5717 nrt) 13700 dwt
147.20 [138.00] x 20.50 x 8.73 m (482.9 [452.8] x 67.3 x 28.6 ft)
1 6-cyl. Afne-Fiat Diesel engine
18000 BHP; 18 kn
by Astilleros Rio Santiago S.A. (Rio Santiago)

1970: completed and delivered to Canumar, p.o.r. Buenos Aires.
Mainly used to transport coal from Hampton Roads to River Plate or ore from Brazil to Plate.
1978 December 22nd: transferred to Alkor Shipping Corp. Monrovia; Liberian flag.
1982: transferred to Shipping Management S.A.M. and re-named *Lucayan Trader*; p.o.r. Nassau.
1986 May 31st: arrived at Gadani Beach, Pakistan to be broken up by Geofman International.

The 1970 *Lodestone* was the second in a class of four sisterships built in Germany for the Ital Pacific Line. They were among the very last traditional freighters to be built and the rapid advance of containerisation (look at her deck cargo in the picture) brought about their quick sale after only a couple of years of service on the Savona-Vancouver route.

Freighters and bulk carriers

PEARLSTONE V
PUERTO ACEVEDO (1973 - 1981)

Bulkcarrier (formerly *Anneliese*, *Twinone*, then *Abuelo Giorgio*)
Builders: Rheinstahl Nordseewerke AG (Emden)
19995 grt (12767 nrt) 28780 dwt
189.00 [180.00] x 25.40 x 10.36 m
(620.1 [590.6] x 83.4 x 34.0 ft)
1 single-acting 6-cyl. Diesel engine; 6340 SHP; 15 kn
by M.A.N. Maschinenfabrik AG (Augsburg)

1961 March 23rd: launched for Kölner Reederei GmbH, Bremen as *Anneliese*.
1972: sold together with her sister ship *Inge* to Twin Shipping Corp. Monrovia and re-named *Twinone*.
1973: sold to Vlasov's Pearlstone Shipping Corp. Monrovia and re-named *Pearlstone*.
1975: re-named *Puerto Acevedo*; management entrusted to Canumar; retained Liberian flag. Mainly used to transport Argentine grain to Japan.
1981: sold to Tarpon Navigation Co. Inc. Monrovia and re-named *Abuelo Giorgio*.
1984 August 19th: arrived at Qingdau to be broken up.

PUERTO ROCCA (1973 - 1980)

Bulkcarrier (formerly *Inge*, *Twintwo*, then *Flag Williams*)
Builders: Rheinstahl Nordseewerke AG (Emden)
19974 grt (12573 nrt) 29643dwt
188.93 [180.00] x 25.40 x 10.36 m
(619.8 [590.6] x 83.4 x 34.0 ft)
1 single-acting 6-cyl. Diesel engine; 9000 BHP; 15 kn
by M.A.N. Maschinenfabrik AG (Augsburg)

1961 August 25th: launched for to Kölner Reederei GmbH, Bremen as *Inge*.
1972: sold together with her sister ship *Anneliese* to Twin Shipping Corp. Monrovia and re-named *Twintwo*.
1973: sold to Vlasov's Diamondstone Shipping Corp. Monrovia, chartered to Canumar Buenos Aires and re-named *Puerto Rocca*.
1977: transferred to Canumar ownership; p.o.r. Buenos Aires. Mainly used to transport coal from Hampton Roads to River Plate and Argentine grain to Europe.
1980 December 31st: sold to Nestos Shipping Co. and re-named *Flag Williams*; p.o.r. Piraeus.
1995 May 30th: laid up in Eleusis Bay, Greece.
1997 October 13th: sold to Indian shipbreakers.

PUERTO MADRYN (1975 - 1977)

Bulkcarrier
(formerly *Aldersgate*, *Silvershore*, then *Danube*)
Builders: Sir James Laing & Sons Ltd (Sunderland)
7689 grt (3996 nrt) 18512 dwt
160.00 [153.92] x 21.34 x 8.99 m
(524.9 [505.0] x 66.7 x 29.5 ft)
1 2-stroke single-acting 4-cyl. Diesel engine
4400 SHP; 12,5 kn
by William Doxford & Sons Ltd (Sunderland)

1960 February 26th: launched as *Aldersgate* for Turnbull, Scott Shipping Co. Ltd, Farnborough.
1969: sold to Bishopsgate Shipping Co. Ltd (Dene Group), Silver Line appointed mangers, re-named *Silvershore*.
1975: after Vlasov Group bought out Dene Line and Silver Line the vessel was transferred to its sub-

The *Puerto Rocca* photographed at the Officine Allestimento e Riparazione Navi berth in Genoa during an overhaul. Note the traditional "V" on her bow.
(Antonio Scrimali, Alpignano)

sidiary Coralstone Shipping Corp., Monrovia and re-named *Puerto Madryn*; management entrusted to Canumar, Buenos Aires. Mainly used to transport grain and ore along South American coasts.
1977: sold to Danube Shipping Co. Ltd, Cyprus and re-named *Danube*; 4 passengers.
1984 April 3rd: arrived at Chittagong, Bangladesh to be broken up.

MARLAND II (1976 - 1979)

LST - WW2 American Tank Landing Ship
(formerly *LST 1150*, *Sutter County*, *Lana*, then *Amal*)
Builders: Chicago Bridge & Iron Co. (Seneca)
3267 grt (2267 nrt) 1575 dwt
93.27 [90.83] x 15.24 x 3.43 m
(306.0 [298.0] x 50.0 x 11.2 ft)
2 12-cyl. Diesel engines; 1700 SHP; 12 kn
by General Motors Corp.

1945 December: built and delivered as *LST 1150*, she was a standard American Tank Landing Ship (L.S.T.); which was originally intended to be used for WW2 landing operations. The *LST 1150* was the third to the last craft in her class to be completed as, starting from 7th December 1941, a total number of 1152 sisterships were built.
1955 July 1st: the LSTs still in service for the U.S. Navy were all re-named after American counties. The *LST 1150* became the *Sutter County*, retaining however her identification number painted on her sides.
1960: re-activated for war service and despatched to Vietnam.
1975: returned to Beuamont, Texas and laid up. Put up for sale. Bought by Vlasov Group, re-named *Lana* and registered for the subsidiary Goldstone Shipping Corp. of Panama. The forward ramp and the other side openings were welded up, and under the command of Capt. Jantzen, the *Lana* crossed the Atlantic to Genoa under its own power. Rebuilt at Cantieri San Giorgio del Porto, Genoa, into a Ro/Ro ferry capable of carrying up to 200 cars or an equivalent number of trailers, 50 of which were part of the permanent equipment on board. She was intended to be used as an harbour ferry for discharging ships at anchor within the Jeddah port area. At the time, in fact, the port suffered a great congestion, with queues of up to 150 ships waiting to berth to unload their cargo. However, when the *Lana* was ready, the new port infrastructure had been completed and thus she became surplus.
1976 April: transferred to Amar Line subsidiary Saudi Maritime Transport Co. Ltd and registered in Jeddah with the new name of *Marland II*. Under the command of Capt. Jachan Berhendt she was employed for livestock trading in the Arabian Gulf, embarking Ethiopian sheep, cows and camels (at the time she was dubbed the "Wa/Wa Ferry", Walk-on/Walk-off, owing to her peculiar cargo).
1979: sold to Abdullaziz Establishment, Jeddah, and renamed *Amal*; used for the lucrative second-hand cars trade between Saudi Arabia and Egypt.
1980 July 24th: the *Amal* ran aground on Falahiyat Reef, in the South Middle Gateway of Jeddah port, with 60 tons of cars on board. After temporary repairs she was drydocked and repaired in Piraeus.
1986 November 10th: whilst laid up at Jeddah Roads she broke her moorings in heavy weather and drifted and stranded. Refloated on the last day of the year.
1987 January 2nd: sank at her moorings owing to leakage of her numerous underwater breaches. In November the *Amal* was sold to a local shipbreaker; demolition was started on 18th December but she definitively sunk on the following 10th January.

The *Marland II* was born as standard WWII Landing Ship, Tank (L.S.T.) built by the U.S.A. and was later employed in the Vietnam War. Here she is seen during a call in Palermo on her delivery voyage from Genoa to Jeddah.
(Antonio Scrimali, Alpignano)

CASTELBRUNO (1934 - 1953)

(formerly *Baharistan*, *Siretul*, *Omega*, then *Castel Bruno*)
Builders: William Gray & Co. Ltd (West Hartlepool)
4327 grt (2654 nrt) 6700 dwt
113.14 [109.73] x 15.54 x 6.92 m
(371.2 [360.0] x 51.0 x 22.7 ft)
1 3Exp Steam Engine; 1500 IHP; 13 kn
by Central Marine Works Ltd (West Hartlepool)

The *Castelbruno* at speed; before joining Sitmar Line in September 1948 she had already served other companies within the Vlasov Group since 1934.

Although she was a minor vessel among the ships owned by the Group, the long life of the cargo-steamer *Castelbruno* covers the earliest, very interesting twenty years of shipping activity by Alexandre Vlasov, telling us much of his own life also.

In September 1934 she was registered, together with the *Prahova* (the eventual *Castelmarino*), under the ownership of Alexandre Vlasov with the name *Siretul*. Her name was later changed to *Omega*, before becoming, in 1948, the Sitmar *Castelbruno*, but her funnel uninterruptedly sported the V for twenty years, from 1934 to 1953, when she was sold for demolition.

Castelbruno (1934 - 1953)

She was laid down on 23rd February 1912 as yard no. 814 at the William Gray & Co. Ltd shipyard of West Hartlepool to the order of F.C. Strick & Co. Ltd.

The steamer was successfully launched with the name *Baharistan* on the following 14th August and, on 17th September, she ran her delivery trials and was delivered to the Anglo-Algerian S.S. Co. Ltd of London, an associated company of Strick's.

On 19th September 1912 the *Baharistan* left the shipyard bound for Swansea, her port of registry.

She was mainly used for the transport of coal between British and Algerian ports; with the typical profile of the one-deck freighters of her time, she had four holds capable of carrying 271,850 cu ft of cargo.

In 1914 the *Baharistan* was bought by the Bucharest-headquartered Romania Prima Societate Nationala de Navigatione to increase their fleet for the transport of coal. All the vessels owned by this company had their names ending in -ul, and the *Baharistan* was transferred to the Romanian flag as the *Siretul*.

The connection between Alexandre Vlasov and the Romanian company had started in the mid-'twenties, when he was chartering its ships to transport coal. Vlasov, in fact, had become one of the most important agents who, on behalf of Romanian industries and electricity plants, imported coal for their needs.

At first a minor partner in the Romania Prima Societate Nationala de Navigatione, Alexandre Vlasov took control in 1933. The company was put into liquidation and its older freighters *Jiul*, *Milcovul* and *Oltul* were sold. Only the *Siretul* and the *Prahova* were kept in service and on 18th September 1934, with the V on their funnels, they became the first vessels of a new company, simply called and registered in Bucharest as Alexandre Vlasov Societate de Navigatione;

The *Castelbruno* started her life in 1912 as the Anglo-Algerian Steamship Co. *Baharistan*.
(Peter Newall, Blandford Forum)

CASTELBRUNO (1934 - 1953)

During the Second World War the eventual *Castelbruno*, at the time called *Omega*, was requisitioned by the British Ministry of War Transport.
(National Maritime Museum, Greenwich)

later a new *Oltul* would join them. The *Siretul* continued to serve the traditional coal route connecting Polish, German and British ports to the Mediterranean and the Black Sea. With the growth of Mr Vlasov's business in other Mediterranean countries, in the mid-'thirties the vessel's voyages started to include calls in many Italian, Greek, Turkish and North African ports.

On 29th March 1938 the *Siretul* cleared her homeport of Braila for the last time after unloading her cargo of coal. During the same year also her fleet-mates *Oltul* and *Prahova* progressively abandoned the coal route to Romania as the business interests of Alexandre Vlasov in that country were almost over and he himself had left Bucharest to move with his family to Milan. This decision stemmed above all from the deterioration of the Romanian political situation: the violent and bloody ascent of the throne by Carol II led in 1938 to the dissolution of the parliament and to the establishment of a dictatorial regime.

Although Italy itself was at the time under Mussolini's dictatorship, Vlasov found in the country a favourable ground for the growing of his business: the Italian peninsula was in a strategic position in the middle of the Mediterranean Sea and Milan was virtually equidistant from the countries with which Vlasov maintained his business relationships.

Castelbruno (1934 - 1953)

On 9th January 1939 the *Siretul* entered Buenos Aires, at the end of her first Atlantic crossing; with the threat of a European conflict at the door, Vlasov, who had economic interests and ships in opposed states, tried to keep his vessels in neutral waters as long as possible.
The *Siretul* and her Romanian fleet-mates gave Vlasov the biggest worries. At the beginning, in fact, Romania was against Germany and Italy, as they supported its revolutionary irredentist movements, but later, in Autumn 1940, Antonescu took full control of the country as a fascist dictator allied to the axis.
From the United States, where he sheltered after Italy's entry into the War, Alexandre Vlasov made a last desperate attempt to save *Siretul* and *Prahova*, transferring them to the Panamanian flag under the ownership of his newly-established Dolphin Steam Ship Corp. (a subsidiary of the New York-headquartered Alvion Steamship Corp.) and giving them the new names of *Omega* and *Tropicus*, respectively.
The change of flag was however not recognised by the Allied authorities and the *Omega*, which on 27th May 1940 took shelter at the Cape Verde Islands, was requisitioned by Great Britain and, on 18th June, brought to Cadiz. Here she remained idle at anchor until 14th January 1941, when she set sail for Lisbon. In the Portuguese port she was again laid up until 26th August 1942 when, registered at Greenock and commissioned by the Ministry of War Transport, she started her War duties under the command of Capt. L. Livingstone. Her management was entrusted to the Navigation & Coal Trade Co. Ltd and thus she was again in some measure under Vlasov control.
She was exclusively used for short trips along the British coast, pushing rarely on as far as Gibraltar, probably owing to the frequent breakdowns to her 30-year old reciprocating engine.
Only in August 1943 the *Omega* re-entered the Mediterranean reaching the Allied base of Augusta, Sicily. She would remain in Italian waters, used as a ferry between the peninsula and the islands of Sardinia and Sicily, until September 1944, when she cleared Naples back again to Great Britain.
On the last day of 1945 she managed with difficulty to reach Dublin, after having been caught in a week-long hurricane, that caused her heavy damage. Repaired, the *Omega* took to the sea again on the following 18th March, but only for a short period.
She went into lay-up at Hull on 13th July and during this period she was again damaged by strong winds which pushed her against the dock and later by a fire on board. De-requisitioned and given back to Vlasov the *Omega* was again repaired and transferred to the ownership of his Buenos Aires-headquartered Canumar, under the Panamanian flag.

Castelbruno (1934 - 1953)

The Omega *portrayed in Montevideo on 23rd June 1948, after her return to Vlasov by the British Ministry of War Tranport and before her transfer to the Italian flag as the* Castelbruno. *(William Schell, Holbrook)*

On 10th June 1947 the *Omega* left Hull bound for Gdynia where she embarked a full load of coal and crossed the Atlantic arriving at New York on the following 12th September, continuing then to Buenos Aires. She remained in service between South American ports until 1st July 1948, when she left Rio de Janeiro heading for Newport News and Genoa.

She arrived in the Ligurian town with her master, Capt. Christenberg, and a crew of 31 men the following 16th August, disembarking 9472 tons of coal. The following day she met in Genoa her old fleet-mate *Tropicus*.

On 9th September both of them were transferred to Sitmar Line and to the Italian flag; the former *Siretul* was re-named *Castelbruno* and the former *Prahova* became the *Castelverde*.

On 26th October 1948, after being drydocked and overhauled, the *Castelbruno* sailed on her first voyage under the Italian flag. Despite 36 years on her back, she served the Sitmar company as a tramp for another four years along the coasts of Europe and North Africa.

Before arriving in La Spezia on 13th May 1953 to be broken up, she made her last voyage from Gdynia to Genoa in Summer 1952. At the time her name had been slightly altered to *Castel Bruno* and, of course, her holds were most appropriately filled to capacity with coal...

CASTELMARINO (1934 - 1953)

(formerly *Prahova, Tropicus, Cloverbrook, Tropicus, Castelverde*)
Builders: Armstrong, Whitworth & Co. Ltd (Newcastle-upon-Tyne)
3597 grt (2129 nrt) 6500 dwt
113.48 [109.91] x 15.54 x 6.80 m
(372.3 [360.6] x 51.0 x 22.3 ft)
1 3Exp Steam Engine; 1700 IHP; 14 kn
by builders

The *Castelbruno* and the *Castelmarino* were the only two freighters owned by Sitmar Line after the Second World War. They had both long careers under the Vlasov banner and the V was first painted on their funnels in 1934.

The *Castelmarino* was launched with the name *Prahova* on 16th September 1921 at the Low Walker yard of Messrs Armstrong, Whitworth & Co. Ltd. On the following 17th March she ran her trials and was subsequently delivered to her owners, Romania Prima Societate Nationala de Navigatione. The brand new *Prahova* joined a fleet which already consisted of four coal freighters, the *Jiul*, the *Milcovul*, the *Oltul* and the *Siretul*, which was to became the Sitmar's *Castelbruno*.

She immediately started a long and busy career as a tramp vessel in European waters but she also made occasional trans-Atlantic crossings to New York under charter. However her logbook shows that from the beginning her main activity was the import of coal for Romania on behalf of the nation's booming industries. Her habitual route was from Polish, German and British ports to Istanbul, Varna

The *Castelmarino* joined the Sitmar Line in 1948 with the name *Castelverde*, which she kept until 1950 when the emigrant carrier *Wooster Victory* took her name.

Castelmarino (1934 - 1953)

(on the Black Sea) and Braila, the Romanian river port on the Danube where she was registered and were she usually disembarked her loads of coal.

In 1917 Alexandre Vlasov settled in Bucharest; in a few years he would became one of the main Romanian coal tycoons, thanks to his outstanding skill as self-made business man. Acting as agent of many local industrial concerns, Vlasov needed several vessels to transport the coal to Romania. Although in early 1928 he made an attempt to run by himself his first vessel (named after his son Boris) he later preferred to charter the *Prahova* and her fleet-mates. It was probably too early for him to enter the shipping business, but at the same time he eventually became the Romania Prima's main customer and later one of its shareholders. The opportunity for him to become a shipowner presented itself in 1933 when, after he signed an agreement with the Polish Skarboferm for the exclusive trade of their coal in the Mediterranean, his financial position became strong enough for him to buy out the Romanian navigation company. One year later the latter company was re-named A. Vlasov S.A. de Navigatione, with headquarters in 47 Akademiei Street, Bucharest.

On 18th September 1934 the *Prahova* was transferred to the new concern with a large white V painted on her black funnel, continuing her service as coal transport in European waters.

She left her habitual coal route for a few trans-Atlantic voyages to South America

A fine view of the *Castelmarino* at speed during World Word Two, when she was in service as the auxiliary transport *Cloverbrook*; note the V on the grey-painted funnel.
(The Mariners' Museum, Newport News)

CASTELMARINO (1934 - 1953)

in September 1936 and between January 1939 and May 1940, when she arrived in Genoa just a few days before Italy entered the War. On 20th May 1940 the *Prahova* left the Ligurian port bound for Hampton Roads, where she remained from 9th June to 19th July, when she resumed her passage, transiting the Panama Canal and trading between Vancouver and Los Angeles. On 17th December she was damaged during a collision with another vessel off San Pedro and put back to Los Angeles to be repaired. It was during her stay in Los Angeles that, on 5th March 1941, she was registered under the Panamanian flag for the Dolphin Steam Ship Co. with the new name of *Tropicus*. The Dolphin Steam Ship Co., a Panamanian subsidiary of Vlasov's newly-established Alvion Steam Ship Corp. (headquartered in New York), was created to move his three Romanian vessels to a neutral flag, to avoid their being seized by the Allied authorities. The change of flag was however not recognised by the American authorities and, after the United States entered the War, on 8th September 1942, while still in lay-up in Los Angeles, the *Tropicus* was duly seized and re-named *Cloverbrook*.

On 25th September 1942, under the management of the American War Shipping Administration, she started a busy career transporting materials between the south-west coast of the United States and Guantanamo Bay, Cuba. On 24th August 1945 she was laid up in Mobile but it was not until January 1947 that she was given back to Vlasov, resuming her former name of *Tropicus* on 22nd February. Keeping the Panamanian flag, she was registered under the ownership of Compañía Argentína de Navegácion de Ultramar (Canumar) and used for general trade in South American waters, exception made for a long voyage to East

In a vain attempt to save the Romanian flagged *Prahova* from Allied requisition, in 1941 Vlasov transferred her to his Panamanian company Dolphin Steamship Co. Inc. with the new name of *Tropicus*. (William Schell, Holbrook)

Castelmarino (1934 - 1953)

Note the Panamanian flag of convenience flying from the stern of the *Tropicus* in a 1947 photograph taken in Montevideo soon after she was given back to Vlasov after her War duties.
(William Schell, Holbrook)

Africa and India, between October 1947 and February 1948.

On the following 17th August the *Tropicus* arrived in Genoa and on 9th September was registered in the maritime district of Genoa for the Sitmar Line with the new name of *Castelverde*. For Sitmar she returned to her old coal route in European waters.

On 16th May 1950 she was re-named *Castelmarino*, and three days later her former name was given to the emigrant carrier *Wooster Victory*.

In Autumn 1951, her last voyage was a long one, which took her to North American waters, pushing as far as Tommy's Arm, Newfoundland.

On 9th August 1952, upon her return to Italy, she was laid up in Trieste and on 5th April 1953 she arrived empty in Genoa, under the command of Capt. Sama and with twentyfive crew members on board. On the following 9th May, under her own power, she made the short trip down the coast to La Spezia, where a local scrapyard started her demolition on 7th July 1953.

CASTELVERDE (1936 - 1942)

(formerly *Inverleith, Sunstone*)
Builders: Harland & Wolff Ltd (Belfast)
6661 grt (4093 nrt)
130.71 [125.97] x 17.00 x 10.51 m
(428.8 [410.0] x 55.8 x 34.5 ft)
1 3Exp Steam Engine; 2796 IHP; 12 kn
by builders

The introduction of the forename "Castel" (Castle) coincided with the foundation of the Sitmar Line. In 1938 Alexandre Vlasov moved with his family from Bucharest to Milan, where his business interests had started four years before, when he established in the Italian city the Sindacato Italiano Combustibili S.A. (known with the acronym Sitcom) for the coal trade.

The Società Italiana Trasporti Marittimi S.A. was founded on 30th April 1938, with a deed by the notary Virgilio Neri of Milan, with an initial capital of 1000 shares of 1000 Italian Liras each, increased during the following November to six million Liras. Alexandre Vlasov appointed his son Boris president of the new company, Carlo di Stefano and Boris Demcenko general managers and Luigi Valazzi managing director. The latter was an old friend of the Vlasov family; they met for the first time when they were living in Odessa. At the time Alexandre Vlasov was sanitary inspector of the local port authority and Valazzi's father was the general manager of the Odessa branch office of the old Italian Sitmar Line (Società Italiana Servizi Marittimi S.A.). This company was absorbed by Lloyd Triestino and closed down in December 1931 and thus it was decided to revive its name, very famous in Mediterranean ports thanks to the company's beautiful vessels, as a good omen for the new Vlasov shipping concern. This however sometimes caused confusion because three other shipping companies, Sitmar (Sbarchi, Imbarchi & Trasporti Marittimi S.A.), Citmar (Compagnia Italiana Trasporti

The *Castelverde* (literally "green castle") was the first vessel owned by Sitmar Line in 1938 and introduced the tradition of the distinctive "Castel" names.
(Museo Navale del Ponente Ligure, Imperia)

Castelverde (1936 - 1942)

Marittimi S.A.) and Trasporti Marittimi Società Anonima were operating at the time in Italy.

To provide an instantaneous fleet for the new Sitmar it was decided to transfer two existing freighters owned by Vlasov's Campden Hill S.S. Co. Ltd of London, the *Pearlstone* and the *Sunstone*, to the new Italian company, with the name of *Castelnuovo* and *Castelverde*, respectively.

The *Castelverde* was, however, the first vessel to enter service under the Italian flag in July 1938 while the other vessel effectively joined Sitmar eight months later as the *Castelbianco*.

The eventual *Castelverde*, yard no. 589 of the famous Belfast shipbuilder Harland & Wolff, was a standard freighter of the War class, "N" type. She had been launched on 26th August 1920 with the name *Inverleith* and delivered to the British Mexican Petroleum Co. Ltd on the following 3rd March. She was the first vessel in a class of five sisterships (*Inverurie*, *Invergoil*, *Inveravon* and *Invergarry*) built for the same owner, all fitted with vertical circular tanks for the transport of petroleum.

After a busy career in European, Central and South American waters she was laid up in Gareloch, Glasgow on 1st September 1930, remaining idle for more than six years.

On 6th October 1936 she became the property of Campden Hill S.S. Co. Ltd and on 6th January 1937 entered the Dutch Wilton-Fijenoord shipyard at Schiedam,

The *Inverleith* just warterborne. In this picture are clearly visible the sharp edges in her hull, typical of many War class vessels. In fact, these "fabricated" ships, as they were called, had been conceived during World War I to be quickly built in improvised yards with poor facilities; to facilitate the hull construction, it was decided to adopt the largest number of flat surfaces possible, later assembled on the slipway, precisely forming the sharp corners. The War class can be thus rightly considered an early example of shipbuilding pre-fabrication. *(Harland & Wolff and Ulster Folk & Transport Museum, Belfast)*

CASTELVERDE (1936 - 1942)

where she was refurbished and transformed into a dry cargo freighter, with the removal of her circular tanks from the holds.

She re-entered service with the new name of *Sunstone* under the Panamanian flag on 1st June 1937, being used both in trans-Atlantic voyages to South America and in Mediterranean waters.

She was transferred to the maritime district of Genoa and to the ownership of Sitmar Line on 26th July 1938, clearing Holy Loch, Scotland, where she had been laid up since the previous 26th March, on 23rd July bound for Danzig. Here she loaded a cargo of coal, arriving for the first time in her new homeport on 24th August. She remained on the coal route between North European and Mediterranean until the start of Second World War, when she was despatched to South America for further trading in neutral waters.

At the beginning of June 1940, a few days before Italy entered the War, she arrived in Naples from her shelter in the Cape Verde Islands and, on the 11th, she was requisitioned by the Italian Navy to be used in convoy trips to North Africa.

Her master remained Sitmar's Captain Innocente Aschero and during the numerous dangerous missions performed with the *Castelverde* he was awarded two War medals for military valour for having twice saved (in September 1940 and in August 1941) with emergency manoeuvres his vessel from air and sea attacks,

The *Inverleith*, eventually Sitmar's first ship, the *Castelverde*, was launched on 26th August 1920 in Belfast by Harland & Wolff.
(Harland & Wolff and Ulster Folk & Transport Museum, Belfast)

Castelverde (1936 - 1942)

A nice photograph of the brand new Inverleith leaving her builder's yard on 3rd March 1921. Note the distinctive flat triangular stern adopted by several hull types of the War class, such as the "N" type to which the Inverleith belonged. (Harland & Wolff and Ulster Folk & Transport Museum, Belfast)

showing "great gallantry, courage and cold blood" in the command of his crew and his steamer.

It was during one of those voyages, at 14.15hrs on 14th December 1942, while steaming 30 miles North East of Tunis, that the *Castelverde* was attacked by the British submarine *Unruffled*: Capt. Aschero managed to avoid four torpedoes, but this time another two hit his vessel. None of the other ships in the convoy could risk stopping to give assistance and while the majority of the crew abandoned the vessel, her master remained on board with a few officers and sailors to check the damage and try to save her. But any effort was made vain when at 16.20hrs a third torpedo hit the *Castelverde* causing a violent fire on board and her subsequent sinking.

It is worth mentioning that crews of requisitioned freighters were civil merchantmen from the original owning company. Capt. Aschero was decorated with a silver medal for military valour for the dedication he showed to his beloved *Castelverde*. He remained with Sitmar Line after the War and would most appropriately become master of the new *Castel Verde*.

CASTELBIANCO (1936 - 1941)

(formerly *Zapala*, *Ovingdean Grange*, *Castelnuovo*, then *Rio Chubut*)
Builders: Lithgows Ltd (Glasgow)
4900 grt (3044 nrt)
125.90 [121.92] x 15.85 x 9.11 m
(413.1 [400.0] x 52.0 x 29.9 ft)
1 3Exp Steam Engine; 2500 IHP; 12 kn
by Rankin & Blackmore (Greenock)

On 30th April 1938 the Sitmar Line was founded in Milan, with a small branch office also in Genoa.

The fleet of the new Italian concern was obtained by putting into liquidation another company of the Vlasov Group, the Campden Hill Steamship Co. Ltd of London, and transferring its steamers *Sunstone* and *Pearlstone* to Sitmar. Vlasov obtained from the port authority of Genoa, in July 1938, a six month provisional certificate of nationality for the *Sunstone*, re-named *Castelverde* and the *Pearlstone*, re-named *Castelnuovo*. However, in July 1938, the *Pearlstone* suffered a severe engine breakdown while on a long charter voyage in Far Eastern waters, and thus she could join Sitmar only in March 1939, but with the name of *Castelbianco*.

She had been launched by Lithgows Ltd of Glasgow as *Zapala* on 9th April 1924 for the Buenos Ayres & Great Southern Railway Co. Ltd, one of the many British-

The *Castelbianco* in Buenos Aires on 1st September 1941, all dressed up in occasion of her sale to the Argentine Flota Mercante de Estado. Note her name on the white stripe of the bow side just over-painted and, under, her new name *Rio Chubut*.
(William Schell, Holbrook)

CASTELBIANCO (1936 - 1941)

The former Castelbianco survived the War and remained in service as the Rio Chubut until 1959. (William Schell, Holbrook)

built and British-owned railways present in Argentina until Perón nationalised them in the 'forties. The *Zapala* and her fleet-mates were entrusted with the trans-Atlantic transport of goods and commodities essential to the running and development of the railways and other freights for their customers.

In early 1935 Houlder Bros. & Co. Ltd took on the task of running the sea services of the Buenos Ayres & Great Southern Railway Co. Ltd, incorporating its fleet. On 19th March 1935 the *Zapala* became Houlder's *Ovingdean Grange*, continuing her trans-Atlantic voyages until 20th October 1936, when, while in Buenos Aires, she was re-sold to Vlasov's Campden Hill Steamship Co. Ltd of London and re-named *Pearlstone*.

In July 1938, on a round-the-world voyage, she was forced to make a long stopover in Adelaide, Australia to undergo many hull and machinery repairs. In fact, during the long trip via Panama Canal which had started in New York on 21st March, she sprang several leaks and her reciprocating steam engine was plagued by continuous breakdowns, compelling the vessel to call in at almost all the

Castelbianco (1936 - 1941)

Pacific islands on the route from Los Angeles to Australia. She resumed her return voyage from Australia only at the end of September and on 27th December 1938 she was laid up in Hull.

Transferred to the ownership of Sitmar Line on 2nd March 1939 and registered in Genoa with the name *Castelbianco*, she arrived from Hull in her new homeport on the following 23rd. She remained in European waters, on the coal route usually served by Vlasov's ships, until the outbreak of the Second World War, when she was employed for two trans-Atlantic voyages in neutral waters. When Italy entered the War, on 10th June 1940, the *Castelbianco* was one of the many Italian vessels which sheltered in Buenos Aires. She remained idle at anchor in the Argentine port until 1st September 1941 when, it being impossible for her to leave the port without the threat of being sunk or seized, she was sold to the Argentine government and, under the management of Flota Mercante de Estado, she resumed service along the South American coast as the *Rio Chubut*.

She continued her service for the Flota Mercante de Estado until 11th May 1959 when, en route from Rio de Janeiro to Buenos Aires, she was caught in a storm and was driven ashore on a reef between Capes Polonio and Santa Maria, near La Paloma. Declared a total loss, her wreck was sold for scrap to F. Sarubbi, Uruguai.

Passenger Ships

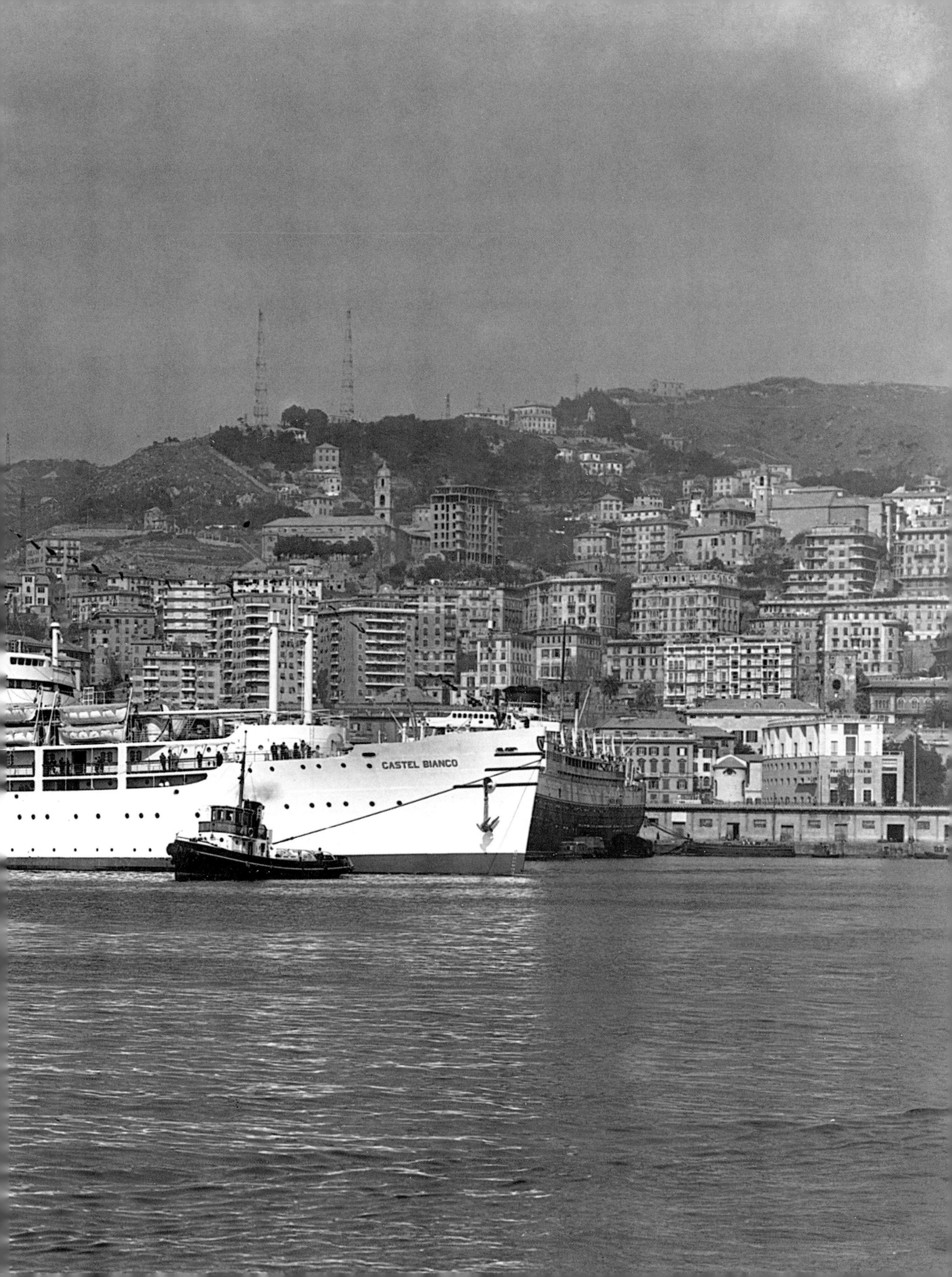
CASTEL BIANCO

CASTEL BIANCO (1947-1957)

(formerly *Vassar Victory*, *Castelbianco*, then *Begoña*)
Builders: Bethlehem Fairfield (Baltimore)
7223 grt (3961 nrt) 3953 dwt; 1953: 10139 grt (5747 nrt)
138.81[133.04] x18.90 x 10.49 m (455.2 [436.5] x 62.0 x 34.2 ft)
1 set of H.P. and L.P. DR geared turbines; 6600 SHP; 17 kn
by Westinghouse Electric Co. (Pittsburg)
1947: 480 emigrants; 48 crew - 1950: 1132 emigrants; 122 crew
1953: 477 tourist class passengers; 717 emigrants

The *Castel Bianco* off Dover after her extensive 1953 rebuilding, based on the plans by the Russian naval architect Vladimir Yourkevitch, Vlasov's friend and the designer of the famous *Normandie*.

The ship which eventually became the Sitmar liner *Castel Bianco* started her life as the *Vassar Victory*, one of four Victory ships bought by Alexandre Vlasov in 1947 for the emigrant service to South America and Australia.

The other three Victory ships to sport the V on their funnels were the *Wooster Victory* (later the *Castel Verde*), the *Patagonia Victory* and the *Argentina Victory*. The two latter vessels, after the impromptu acquisition of a larger C3 vessel (the eventual *Fairsea*) with hull and machinery of superior characteristics, were sold in August 1948 to the Holland America Line, becoming the *Arensdijk* and the *Akkrumdijk*, respectively. The *Vassar Victory* and *Wooster Victory* would, on the contrary, be transformed into emergency emigrant ships, being two of the 97 Victories originally fitted out as troop transports.

Castel Bianco (1947-1957)

The *Vassar Victory* was laid down on 19th March 1945 in Baltimore (her port of registry) by the Bethlehem Fairfield Shipyard. She was launched on the following 3rd May, arriving in New York at the end of the month.

On 6th June she was officially handed over on Lend-Lease terms to Great Britain and her fitting out as a troop transport was started. Actually, when the ship was ready the War was over and thus she was returned to the Americans who used her to repatriate their soldiers from Europe. On 20th September 1945 she sailed on her maiden crossing to Seine Est and Le Havre. She made a total of eight round voyages between European and North American ports, completing the last one in Baltimore on 4th April 1947.

Sold to Vlasov's Compañía Argentína de Navegación de Ultramar (Canumar) and transferred to the Panamanian flag, on 29th May the *Vassar Victory* left Boston bound for the Adriatic port of Ancona, where she arrived on 19th July, after calling at Antwerp, Charleston and Gibraltar. The following day she was re-named *Castelbianco* and transferred to the Italian flag.

On 21st July 1947 in Geneva the IRO (International Refugee Organisation) started its Mass Resettlement Scheme allocating the transport of new settlers and displaced persons to private shipowners including Alexandre Vlasov who had actually bought his Victories for this purpose.

Before being transformed into an emigrant vessel the *Castelbianco* was used for one voyage to Far East as a freighter; the photographs shows her in this guise during her March 1948 call in Cape Town harbour.
(Alex Duncan, Gravesend)

Castel Bianco (1947-1957)

However, in her first months with the Vlasov Group the *Castelbianco* was employed as a cargo vessel. She left Ancona on 16th August 1947 for a long, eventful voyage around the World, during which she had two collisions with other freighters, a grounding and was caught in a severe storm which forced the ship to make a two-month stopover in Sydney for repairs, during which spartan accommodation for 480 passengers was also fitted. She called at 23 ports, among them New York, St. Thomas, Puerto Belgrano, Colombo, Madras, Palembang, Sydney, Aden and Baltimore. At the conclusion of this voyage she arrived in Genoa for the first time on the morning of 14th October 1948 under the command of Capt. Innocente Aschero and with only her 44 crew members on board. She left 5 days later fully booked with I.R.O. assisted passengers for Sydney where she arrived on 19th November 1948. On the return leg of the voyage she made a special call at Shanghai where she was filled to capacity with Russian Jews, who were fleeing from China.

Three similar emergency trips were made in March, June and October 1950, when the *Castelbianco* embarked in Djakarta Dutch citizens leaving their colony after its proclamation of independence and brought them to Rotterdam and Amsterdam.

The *Castelbianco* remained in regular service on the Genoa-Naples-Melbourne-

A 1950 ceremony when crossing the line on board the *Castelbianco*.
(Rodolfo Potenzoni, Genoa)

In September 1950 the *Castelbianco*'s capacity was increased to 1132 emigrants; on this occasion her hull was repainted white.
(Antonio Scrimali, Alpignano)

Castel Bianco (1947-1957)

Sydney route until 19th April 1950, when she sailed bound for Australia from Bremerhaven, her new European terminus. On 13th July 1950 the steamer arrived in Genoa from Bremerhaven and she was docked at Calata Sanità fitting out quay where she underwent the first of two radical transformations into a passenger liner. A row of portholes was opened in her hull, now repainted in white, the small house behind the stack was enlarged to fit a social lounge, 18 lifeboats with their davits were added to the four already on board. Inside steerage accommodation for 1132 passengers was fitted and the crew was increased to 122 men. On 4th September 1950 the *Castelbianco* was ready to take to the sea again on the Bremerhaven-Sydney route, now on a more regular basis of 5 round-trips per year in accordance with the IRO schedule. She was at the time under the command of Romeo Devoto, who in the 'thirties had been the skipper of the famous Guglielmo Marconi's yacht *Elettra*.

On 30th August 1952 the *Castelbianco*, back from Australia, reached Trieste and the following day entered the Cantieri Riuniti dell'Adriatico yard at Monfalcone to undergo the final and most extensive rebuilding of her career. A new upper shelter deck (Saloon Deck) was built at the level of the old forecastle and a new large 2-deck superstructure (Boat Deck and Sun Deck), surmounted by a central house, added. The masts and the cargo booms were removed and the wheelhouse, surmounted by a small mast, raised one deck. Many portholes were opened in her sides to create new cabins (111 outside and 175 inside) in the space previously occupied by cargo holds; there were now on board 1194 berths, 717 in 24 dormitories with a minimum of 20 and a maximum of 42 beds each.

On the lower B Deck were fitted the dormitories while on the upper A Deck and Promenade Deck there were the majority of the cabins, among them 56 with private facilities. The Saloon Deck housed the two dining rooms, the ladies room, the children's room and the writing room, and was encircled by the promenade. The upper Boat Deck accommodated eight cabins, a ball room, a bar lounge and, aft, a lido with open-air swimming pool.

The lido with swimming pool and the main lounge after the 1953 refit.

Castel Bianco (1947-1957)

In this photograph taken on 27th September 1953 during her docking at the Bremerhaven Columbus quay, the collision damage sustained a few days before in the St. Lawrence river by the *Castel Bianco* is clearly visible.
(Arnold Kludas, Grünendeich)

When the steamship, with her name amended to *Castel Bianco*, left the yard and entered Genoa in the evening of 3rd March 1953 she was hardly recognisable, thanks also to a new funnel casing. It was said at the time that the superstructure was too big and top-heavy and this would explain why the ship was a famous roller, even in smooth seas or when moored; and actually, when the *Castel Verde* was rebuilt soon afterwards on the same plans used for the *Castel Bianco*, it was finally decided to add one deck less on the upperworks. On 7th March 1953 (see page 60), a beautiful Mediterranean Spring day, the *Castel Bianco* sounded her siren and cast off all lines from the Maritime Station of Ponte dei Mille and, all dressed up in her smart white livery, she inaugurated the new Genoa Curaçao La Guaira route for Sitmar.

She had an immediate success and, in fact, after her maiden voyage to Sydney, the Sitmar's flagship *Castel Felice* became the *Castel Bianco*'s running-mate on the Central America route.

The *Castel Bianco* maintained the regular service to the Caribbean until her transfer in December 1956 to the New York run, although before that she had made two occasional voyages to Quebec with Hungarian emigrants in Fall 1953 and a final visit to Sydney in November 1956.

She left New York on what was to be her last crossing under the Italian flag on 5th February 1957, under the command of Capt. Giuseppe Mortola. She reached the Ligurian port on the following 17th where she was laid up and put up for sale. The technical representatives of the Compañía Trasatlántica Española came to Genoa on 8th March, carefully visited the ship while in drydock, and confirmed the decision to buy her. On 17th March 1957, with the new name of *Begoña* (after the patroness saint of the city of Bilbao), she left Genoa for Barcelona where a revamping of her accommodation was carried out, eliminating the dormitories and thus reducing the total number of Tourist passengers to 830. One month later the *Castel Verde* would also be sold to the Spanish company, becoming the *Montserrat*.

In May 1957 the *Begoña* was again in service with the first voyage under the Spanish flag to Sydney, where she arrived on 20th June.

One year later she was transferred with her sistership *Montserrat* to the Europe-Venezuela run which she would successfully maintain for a further 17 years.

Her end came at the end of 1974; on 27th September she left Southampton for one of her crossings to the West Indies but soon her old turbines showed signs of malfunction. The steamer headed to Tenerife for repairs and, on 4th October re-

Castel Bianco (1947-1957)

started her journey. But on 10th October, off the coast of Barbados, the main engine broke down and she remained adrift for a week with 800 passengers on board, until on the following 17th the German tug *Oceanic* towed her to Bridgetown. As the *Begoña* was already approaching her programmed retirement (the *Montserrat* was withdrawn from service in February 1973) it was decided not to carry out any repairs. Towed to Castellon on Christmas Eve, she was demolished by the local Spanish shipbreakers in the early months of 1975.

The *Begoña* leaves Southampton docks assisted by the tug *Brockenhurst*. (National Maritime Museum, Greenwich)

CASTEL VERDE (1947-1957)

(formerly *Wooster Victory*, then *Montserrat*)
Builders: California S.B. Corp. (Los Angeles)
8254 grt (4698 nrt) 10753 dwt; 1953: 9001 grt (4758 nrt) 4375 dwt
138.68 [133.04] x18.90 x 10.49 m
(455.0 [436.5] x 62.0 x 34.2 ft)
1 H.P. and L.P. DR geared turbines; 6600 SHP; 17 kn
by Allis Chalmers Manufacturing Co. (Millwaukee)
1947: 890 emigrants; 220 crew
1950: 24 cabin class passengers; 890 emigrants; 260 crew
1953: 455 tourist class passengers; 578 emigrants; 275 crew

The third *Castel Verde* in Sitmar history was a sister ship to the second *Castelbianco* and, like the latter, she was born as a War World Two emergency freighter of the famous Victory class.

In Italy, at the end of the hostilities, the difficult economic situation, the lack of steel and the destruction of many industrial plants, brought Italian shipowners and shipyards to develop a great ingenuity in the recovery and in the transformation of War-built freighters into passenger ships to cope with the immediate demands of the huge new flood of migration.

Some of them were fine transformations and were acclaimed at international level: the Italian Line's Navigatori Class, the Costa Line's *Andrea C.* (a former Ocean class freighter) and the Lauro Line's *Roma* and *Sydney* (former C3 hulls).

On 24th May 1950 the Castel Verde left Naples on her first voyage to Buenos Aires. (Admeto Verde, Naples)

CASTEL VERDE (1947-1957)

In the rebuilding of the Italian Merchant Fleet the Vlasov Group played an important role, placing in service the *Fairsea* and the *Fairsky* (former American C3 vessels) and the *Castel Bianco* and *Castel Verde*.

The latter was laid down on 9th February 1945 and launched on the following 2nd April as *Wooster Victory* by California Shipbuilding Corporation of Los Angeles (her port of registry) and, completed as a troop transport and placed under the management of General Steamship Corporation, on 10th May immediately sent to the Pacific. However, when she made her first visit to Melbourne, on 29th May 1945, the conflict was almost over. She was then despatched to South African waters and later made six round Atlantic crossings to France repatriating Allied soldiers. On 4th May 1946 the *Wooster Victory* concluded her service for the U.S. War Shipping Administration in New York and, after a short period spent at Hampton Roads, on 21st August 1946 she was laid up in the James River Reserve Fleet and put up for sale.

The *Wooster Victory* was bought with three other sister ships (*Lake Charles Victory, Rollins Victory* and *Vassar Victory*, the eventual *Castelbianco*) and three Liberties (*Samlouis, Harry Percy* and *Susan Colby*) by Alexandre Vlasov in January 1947 and registered under the Panamanian flag and the ownership of his Buenos Aires-headquartered Canumar. For the time being the *Wooster Victory* retained her original name. She left her anchorage in the James River on 17th February 1948 and, after some reclassification works and general overhaul carried out in Baltimore, she set sail on 22nd July from New York and, after loading cargo in South American ports, she crossed the Atlantic bound for Genoa.

On 6th August, after having been transferred to the Alvion S.S. Corporation, she weighed anchor from the Ligurian port bound for Australia for her first voyage under the auspices of the International Refugee Organisation, who kept on board its escort officer, one medical doctor and four nurses to give assistance to the new

The *Castel Verde* started her life as the U.S. Troop Transport *Wooster Victory*; her funnel sports the initials of her managers, General Steamship Corporation.
(Peabody & Essex Museum, Salem)

Castel Verde (1947-1957)

settlers during the voyage. One month later she disembarked her first passengers in Sydney, 892 refugees and displaced persons. Each of them had paid for the passage a symbolic price of 5 Pounds for the ticket, inclusive of a bonus to obtain cigarettes and drinks from the ship's bar. In November 1948 while she was leaving Melbourne bound for Europe, she received via radio the order to alter course and, on behalf of U.N.O., to head to Manila to take on board a pilot, and wait for the authorisation to enter Shanghai and embark White Russians and Jews running away from the new Maoist Regime in China. She left Shanghai on Christmas Eve and one week later her sister ship *Castelbianco* reached the Chinese harbour for the same purpose.

Like her sister ship, the *Wooster Victory* underwent two major rebuildings during her career under the Vlasov banner; the first one was carried out in Genoa in 1950. She arrived in the Ligurian town on 2nd January and berthed at Calata San Lazzaro quay. Two rows of portholes were opened in her hull and new cabins and dormitories for 914 passengers were fitted; a new galley was installed and the house aft the engine casing was enlarged to fit an ample new dining room, used also as social lounge. Outside the appearance of the ship was only slightly altered: the hull was painted white and a new ellipsoidal funnel casing, shorter but larger than the previous cylindrical one, fit-

Towards a new life: Jewish refugees from Shanghai dance on the quay after disembarking the *Wooster Victory* in Cape Town.
(*South African Library, Cape Town*)

A 1949 view of the emigrant ship *Wooster Victory* in Cape Town harbour; note the post-war funnel colours adopted by Vlasov, a blue V on a yellow background instead of the much cheaper to maintain white V on a black background used in the pre-war era.
(*Alex Duncan, Gravesend*)

Castel Verde (1947-1957)

ted. Lifeboats, masts and cargo-booms remained unaltered. On 16th May 1950 Sitmar's freighter *Castelverde* was re-named *Castelmarino* and thus the *Wooster Victory* took her name, slightly modified to *Castel Verde*, on the 19th. She re-entered service from Genoa on 22nd May 1950 with 914 passengers on board bound for Buenos Aires, her new route, but still flying the Panamanian flag. She was effectively transferred from Alvion to Sitmar and to the Italian flag in September, her port of registry being Rome.

She would maintain her regular service on the Genoa, Naples, Palermo, Lisbon, Funchal, Las Palmas, Rio de Janeiro, Santos, Montevideo, Buenos Aires route for another four years, completing also four occasional voyages to Australia, in July 1953, April, June and October 1954. At the time the vessel was advertised on posters and in local newspapers as "C. Verde", in order to capitalise on the still high reputation gained by the famous pre-War Lloyd Sabaudo liner *Conte Verde*, whose memory was still synonymous with high class service among South American people.

One month after the *Castel Bianco* was greatly upgraded at the Cantieri Riuniti dell'Adriatico Monfalcone shipyard and before being transferred to the Central America route, the *Castel Verde* was also rebuilt on the same lines, although her superstructure was given one deck less. Back from South America, she entered the Muggiano works (La Spezia) on 25th April 1953 and left on the following 2nd June with a totally new look. She could now embark 1033 passengers, 455 in 124 cabins and the remainder in 32 dormitories.

On 7th July 1954, back from Australia, she started a new regular line service from Genoa to the Caribbean, via Vigo, Lisbon, Madeira and Tenerife, completing the crossing at Curaçao on the 24th of the same month.

She was partnered on this route by her near-sister *Castel Bianco* in transporting Italian and Spanish emigrants to Venezuela, while on the return to Europe she called at Kingston to embark Jamaicans on the "assisted passages" scheme migrating to Great Britain; Mr Vlasov had in fact

A standard outside two berth cabin and the dining room on board the *Castel Verde* after the 1953 refit.

71

CASTEL VERDE (1947-1957)

The Castel Verde *showing her new look after the 1953 rebuilding (Arnold Kludas, Grünendeich)*

signed a contract with the Jamaican Government which at the time paid a contribution of 125 Pounds Sterling per passenger.

In the evening of 28th March 1957 the *Castel Verde* entered Genoa under the command of Capt. Maiocco and with 201 crew members at the end of what was to be her last voyage for the Sitmar Line. Sold to the Spanish Line with the *Castel Bianco* (renamed *Begoña*) for 9-million dollars, she was renamed *Montserrat* on 16th April and, five days later, drydocked for a general overhaul.

She left Genoa flying the Spanish flag on 23rd April 1957 and, after a revamping of her 825-berth accommodation at the Astilleros Espanoles of Barcelona, resumed service on the Caribbean route, maintaining the same itinerary she had previously followed for Vlasov.

The *Montserrat* was also used to bring West Indian migrants to England and made also a last voyage to Sydney in 1959. The latter proved to be one of the most difficult voyages of her career: she left Naples on 6th May but soon her boilers showed signs of malfunction and she had to proceed at reduced speed. She was forced to make a two-week stopover in Colombo for the most urgent repairs, arriving at Fremantle only on 29th June. Here her lifeboats failed the inspection by the local port authorities who denied her permission to continue the voyage to Melbourne and Sydney. Her passengers were forced to disembark and after repair to her lifeboats, *Montserrat* was finally allowed to sail from Fremantle on 8th July.

In 1962 she was fully air-conditioned and the number of passengers was reduced

After a few voyages during which she retained the Sitmar white hull with blue band, the hull of the Montserrat *was painted in the traditional black livery adopted by Spanish Line vessels. (World Ship Society, Kendal)*

Castel Verde (1947-1957)

During their last period of service both *Montserrat* and *Begoña* returned to white hulls; the extended funnel casing of *Montserrat* shows also the new logo adopted by the company, its blue flag with a central white circle on a yellow background.
(Francis Palmer photo, Steamship Historical Society of America, Baltimore)

to 708 berths in two classes. In the process also the old all-black funnel casing was heightened and painted in yellow with the company's flag in the middle.

In August 1970 her turbine set broke down in mid-Atlantic and her passengers were transferred to the *Begoña* while the *Montserrat* was towed to Willemstad for repairs.

She re-entered service the following December but her machinery continued to cause problems. After two years she was finally withdrawn from service and on 3rd March 1973 she was delivered at Castellon to I.M. Varella Davalillo to be broken up.

FAIRSEA (1949-1969)

(formerly *Rio De La Plata*, **HMS Charger**)
Builders: Sun S.B. & D.D. Co. (Chester)
11678 grt (5800 nrt) 4600 dwt; 1955: 13433 grt (7606 nrt) 5319 dwt
149.96 [141.57] x 21.18 x 7.31 m
(492.0 [464.5] x 69.5 x 24.0 ft)
2 single-acting 2-stroke 12-cyl. Doxford geared Diesels; 8500 SHP; 16.5 kn
by builders
1950: 1800 emigrants; 210 crew
1955: 1620 passengers; 237 crew
1961: 1212 one-class passengers; 240 crew

In 1936 the United States Congress passed a new Merchant Marine Act which was to make important changes to the previous system of state subsidies given to American shipowners and shipbuilders, by substituting the newly constituted Federal Maritime Commission for the old Shipping Board, founded in 1916.

One of the main targets of the commission was the renewal of the national fleet of freighters, which at that time mainly consisted of old vessels built as emergency transports during the First World War. For this purpose they planned the C3 type, a hull of approximately 10,000 dead-weight tons and a displacement of about 15,000 t, suitable for the transport of general cargo and without route limitation. In July 1938 the preliminary general plans and particulars of this standard freighter were published in the famous American magazine Marine Engineering with a public invitation to contact the technical department of the Maritime Commission with suggestions and alterations in order to improve the vessel and make it more suitable to owners needs. As a result of this enquiry it

The *Fairsea* during her maiden year of service for Vlasov.

FAIRSEA (1949-1969)

At the end of the hostilities the *Charger* was decommissioned and went into lay-up on the Hudson river.
(The Mariners' Museum, Newport News)

Before putting her up for sale, the American Maritime Commission arranged the removal of *Charger*'s armament and flight deck.
(The Mariners' Museum, Newport News)

became evident that many companies required the vessels to be fitted with limited accommodation for the carriage of passengers. Following co-operation with the design departments of many American owners and shipyards, the project, under the guidance of the senior Naval Architects G. Sharp and V.M. Friede, came to fruition in Summer 1940, when the Moore-McCormack Lines' *Sea Fox*, the first C3 vessel, entered service.

The ship destined to become Vlasov's *Fairsea*, was part of the group of the twentyfour C3 combi-ships (known as the C3-P type) and specifically one of the four units of the Rio class of the Moore-McCormack Lines, all of them engined by a pair of 9,000 BHP Doxford Diesels acting on a single screw and giving the vessel a service speed of 16 knots. She was laid down as Yard no. 188 on 19th January 1940 at Chester, Pennsylvania in the yard of the Sun Shipbuilding & Drydock Co. and launched on 1st March 1941 as *Rio De La Plata* by Señora Felipe Espil, wife of the Argentine Ambassador to the United States. The ceremony was a quite solemn affair with the presence of North and South American dignitaries, and was broadcasted live in the United States and in Argentina.

Moore-McCormack Lines intended to outfit their four new liners for their New York-Rio de Janeiro

Fairsea (1949-1969)

Before her transfer to Cantieri Navali del Tirreno in Genoa for the final fitting out, the Bethlehem shipyard in Hoboken built the new upperworks of the eventual *Fairsea*, which still sports her original name *Charger* on the stern.
(Arnold Kludas, Grünendeich)

route with 76 outside first class cabins for 196 passengers. All the cabins were to have private facilities and 20 of them would be of the luxury type, with a private veranda.

The *Rio De La Plata* was to enter service in February 1941 together with her sisterships *Rio Hudson*, *Rio Parana* and *Rio de Janeiro* but the War altered the fate of all of them.

They were all requisitioned in May 1941 under the Lend-Lease agreement with Great Britain and were commissioned into the Royal Navy after completion as

In December 1953 the *Fairsea* was given a new funnel casing and a tripod on the bridge roof replaced her masts.

76

Fairsea (1949-1969)

Aircraft Escort Carriers. The *Rio De La Plata* was re-named H.M.S. *Charger* (BAVG-4) and put under the command of Capt. George Abel-Smith who supervised her transformation at the Newport News Shipbuilding and Drydock Co. Here the partially-built upperworks were removed, a flight deck installed and a hangar for 18 aeroplanes fitted into her former cargo holds.

However the *Charger* was not commissioned into Royal Navy as, upon completion on 4th October 1941, she was transferred to the U.S. Navy which, on 24th January 1942 commissioned her as AVG-30 retaining the British name. On the following 3rd March she commenced her War duties under the command of Capt. T. L. Sprague.

Charger's area of operation throughout the conflict would be Chesapeake Bay where she was used as a training vessel for pilots and ships crews in carrier operation.

She was reclassified ACV-30 on 20th August 1942 and CVE-30 on 15th July 1943, leaving Chesapeake Bay only twice: in October 1942 for a ferry crossing to Bermuda and in September 1945 for a voyage to Guantanamo Bay, Cuba. Decommissioned in New York on 15th March 1946, she was laid up on the Hudson River and later that year her armament and flight deck were removed. On 30th January 1947, after Moore-McCormack Lines refused to accept re-delivery of the ship, she was placed in the hands of the Maritime Commission and put up for sale.

The *Charger* remained in lay-up for another two years until she attracted the attention of Alexandre Vlasov, who had just signed a contract with the

A 25th May 1955 photograph of the *Fairsea*'s promenade during her extensive refitting in Monfalcone.
(Associazione Marinara Aldebaran, Trieste)

Fairsea (1949-1969)

International Refugee Organisation (IRO) to transport emigrants and displaced persons. On 8th May 1949 the vessel became the property of the Alvion Steamship Corporation of Panama and was placed under the management of Sitmar Line. The sale agreement specified that some of the conversion should be carried out in an American shipyard and thus the former *Rio De La Plata* was towed to the nearby Bethlehem Steel yard in Hoboken. Here, under the supervision of the naval architect Dario Rivera, the hull and the machinery were overhauled; a two-deck superstructure surmounted by a bridge and funnel were fitted and masts, kingposts and derricks for cargo handling were added.

With the new name of *Fairsea* she sailed under her own power for Genoa where, at the local Cantieri Riuniti del Tirreno, her spartan public lounges and dormitories for 1800 steerage passengers were added.

On 3rd December 1949 the *Fairsea* set sail from Genoa on her maiden voyage to Australia. Upon her arrival in Fremantle she had a special welcome, because she disembarked the 50,000th post-War immigrant arriving in the continent. The vessel later proceeded to Sydney, where she arrived on the 30th of the same month. She spent the following two years in regular service between Europe and Australia; outward voyages were usually filled to capacity but she ran light on the homeward run. For this reason, in February 1952 Sitmar opened an agency in Sydney, Navcot Australia Corporation, to offer low-priced passages on the return leg of the voyages, with many intermediate calls en route to Europe.

Between 30th April and 18th September 1953 the *Fairsea* was moved from her customary itinerary to make six round crossings between Bremerhaven and Quebec.

One of the dining rooms after the 1955 refit; note the modern sprinklers (introduced by Solas 1949) and the air conditioning diffusers on the ceiling.
(Associazione Marinara Aldebaran, Trieste)

Fairsea (1949-1969)

In November 1953, while the ship was in Melbourne, and some modification works were being carried out to the diesels feeding system, there was an explosion and a subsequent fire in the engine room. It was extinguished with no casualties and the ship was able to return to Genoa where the works were completed. On this occasion the original funnel was replaced by one of more modern design and the two masts replaced by a tripod fitted on the bridge roof.

On 1st April 1955 the liner entered the Cantieri Riuniti Dell'Adriatico yard of Monfalcone for the most important refit of her career. The boat deck was extended fore and aft and the superstructure raised one deck in order to include a new house containing a First Class section with ten 4-berth cabins, a lounge and a dining room: this increased the gross tonnage to 13,433 t. The already existing 286 Tourist Class cabins for 1620 passengers were refurbished as well as the accommodation for the 28 officers and the 209 members of the crew. An air conditioning plant was installed and the fire-fighting sprinkler system extended to all cabins and lounges. The electrical wiring, lighting, fire doors and the galley equipment were all renewed. The work, supervisioned by Boris Vlasov in person, was done in a record time to fulfil the owner's commitment with the I.C.E.M. (International Committee for European Migration) for voyages to Australia. The conversion was considered one of the most successful carried out in that period by C.R.D.A. following the *Conte Biancamano* and the *Homeric* refits.

On 1st July 1955 the *Fairsea*, saluted by the sirens of the Navy Second Division then present in the harbour, left the Maritime Station of Trieste with 1424 emigrants bound for Sydney. Among her passengers was also Boris Vlasov: he arrived suddenly at the Maritime Station when the ship had already let go her lines and

Wellington, 7th January 1967: the *Fairsea*, during her last period of service, is seen on one of the few cruises she made during her life. *(Victor Young & Len Sawyer, Wellington)*

Fairsea (1949-1969)

A fine 1967 photograph of the Fairsea at speed in the Solent. (Ambrose Greenway, London)

was slowly leaving the pier and thus she had to dock again to embark the owner. In the following December she replaced the Cunard-White Star liner *Georgic* (on charter to the Australian Government since January 1949 for assisted emigrant passages) and her European terminus port became Southampton. It was a very brilliant move by Alexandre Vlasov and the *Fairsea* became the first non-British ship to be chartered for the transport of British emigrants to Australia.

On 20th June 1957 the *Fairsea* returned to the North Atlantic run, departing Rotterdam for Quebec via Le Havre, and in August she started the first of three round voyages to New York, under charter to the American Overseas Charters and Shipping Co.

At the beginning of 1958, the Vlasov Group had signed another four-year contract with the Commonwealth Government, and the *Fairsea* was transferred to the ownership of Sitmar and registered in the Maritime District of Rome; in April that year the vessel entered service on the Bremerhaven-Australia route, via Panama Canal on the return leg, together with her running mates *Castel Felice* and *Fairsky*. In December 1961 her 8-, 10- and 12-berth dormitories were eliminated and the passenger capacity reduced to 1212. After the charter to the

Fairsea (1949-1969)

Australian Government lapsed, the motorship continued on the route and on 7th July 1966 left Sydney for the first of the very few cruises made in her career, calling at Hayman Island, Brisbane and Melbourne. Two years later she was again transferred to the Liberian flag, under the ownership of Sitmar's subsidiary Passenger Liner Services Inc. of Monrovia.

The planned withdrawal of the ship in 1970 was hastened by an accident which occurred while she was en route from Papeete to Balboa. On 23rd January 1969 a fire broke out in her engine room leaving the ship without power. The Shaw, Savill liner *Athenic* responded to the *Fairsea*'s SOS, reached the vessel on the 25th and stood by until the arrival of the tug *R. Pace*. After the tow line was secured the *Athenic* left the *Fairsea* and resumed her passage. Unfortunately during the evening of the same day, the tug broke down. The *Fairsea*'s engineers went on board the *R. Pace* and did their best to partially restore the power on it, but it was insufficient to tow their liner. The *Fairsea*'s situation became very serious: the galleys, the air conditioning, the potable water distillation plant and the toilettes for her 985 passengers and 240 crew members had been out of action for four days. This fact placed a severe strain on her master, Capt. Ciro Cardia: in despair he retired to his room and committed suicide.

It was not until 29th January that the Lykes Brothers' freighter *Louise Lykes* was able to take the *Fairsea* in tow and, behind her, the tug *R. Pace*. The convoy arrived at Balboa in the early afternoon of 3rd February where the *Fairsea* was laid up and put up for sale "as is, where is" because of the damage to her machinery, the Doxford Diesel engines being too old and costly to be worth repairing.

She was sold for 300,000 US$ to the La Spezia demolition yards; on 9th July 1969 the Italian tug *Vortice* commenced towing the *Fairsea* to the shipbreakers yard where she arrived the following month and was broken-up.

CASTEL NEVOSO (1949-1968)

(formerly *Friesenland*, *Fairsky*, then *Argentina Refeer*)
Builders: Howaldtswerke AG (Kiel)
3828 grt (1642 nrt) 2987 dwt; 141.00 [129.00] x 16.50 x 5.40 m
(462.6 [423.2] x 54.1 x 27.0 ft)
2 9-cyl. Diesel engines; 5800 SHP; 16 kn
by M.A.N., Maschinenfabrik Augsburg-Nürnberg AG (Augsburg)
24 passengers; 44 crew

Of all the ships in the Vlasov Group fleet, the *Castel Nevoso* was one of the most interesting, with a long and eventful career, starting in the 'thirties as the seadrome *Friesenland* of the Lufthansa airline.

In the inter-War period Germany played an important role in the development of trans-Atlantic flight; famous airships, such as the *Graf Zeppelin* or the ill-fated *Hindenburg* are well known examples.

In order to speed up the delivery of the trans-Atlantic mail, in the early 'thirties the Norddeutscher Lloyd liners *Bremen* and *Europa* had their own seaplanes; they were launched by a catapult when the liners were one and a half days out of New York or Bremerhaven, saving in this way 24 hours in the delivery of the mail.

A few years later the Deutsche Lufthansa AG developed their own system (sponsored by the German Government) for the management of the trans-Atlantic air mail. As the main problem was the inability of the airplanes to carry sufficient

*The Castel Nevoso departing Genoa in June 1966 on one of her regular sailings to Chisimaiu.
(Victor Young & Len Sawyer, Wellington)*

CASTEL NEVOSO (1949-1968)

The *Castel Nevoso* started her life as the seadrome *Friesenland* of the German Lufthansa airline. *(Lufthansa AG Historical Archives, Munich)*

fuel to cross the Atlantic, they transformed two freighters, the steamer *Westfalen* and the motorship *Schwabenland*, into replenishment vessels for their long-range seaplanes.

These vessels were positioned along the route followed by their seaplanes, which, after landing nearby, were refuelled before continuing their voyage.

These experiments proved successful and Lufthansa ordered two larger, purpose-built motorvessels, the *Ostmark* and the *Friesenland*.

The latter, yard no. 755 at Howaldtswerke AG, was launched in Kiel on 23rd March 1937. She left her builders on 13th May 1937 to run her sea trials in the North Sea, heading then for Hamburg where the specially-built crane and the catapult were fitted on board.

The *Friesenland* was a real floating airport, with a sheltered hangar acting also as workshop for the repairs of seaplanes, a wireless apparatus to keep a constant radio contact with the planes and sophisticated meteorological equipment for the continuous despatch of weather-reports.

The huge stern crane was able to lift large seaplanes weighing up to 17 tons; the vessel could accommodate two of them, one in the hangar and one on the catapult.

The crew accommodation, considering the long periods the vessel would spend in open waters without calls, was particularly comfortable: large cabins, many with private facilities, a gymnasium, a social lounge and a large library were among the amenities on board.

Of particular note was the flat transom stern of the vessel: now a very common solution in shipbuilding, the *Friesenland*'s stern was one of the very first of its kind. The reason was the necessity to fix behind it a large floating net that, towed

CASTEL NEVOSO (1949-1968)

In Summer 1937 the *Friesenland* underwent a series of trials in the North Sea to test the lifting on board and the catapulting-off of the large Lufthansa trans-Atlantic seaplanes.
(Enrico Repetto, Genoa)

by the vessel, created an area of calm sea to facilitate the seaplanes landing.

After a series of trials in the North Sea during which the *Friesenland* lifted on board, refuelled and re-launched Lufthansa seaplanes, she left Bremerhaven on her first mission on 15th August 1937. She was positioned a few miles west of the Azores, remaining there until the end of September, when she sailed for New York for her own refuelling.

Her second mission was between February and June 1938, this time being positioned along the South Atlantic route, using the ports of Santos and Pernambuco as bases.

Back to Germany via New York, she sailed on her third mission on 30th October 1938, again bound for the South Atlantic.

With the worsening of the International political situation and the prospect of War, the *Friesenland* set sail from Pernambuco on 3rd February 1939 bound for Las Palmas where she remained until 3rd August, when she resumed her voyage to Bremen, arriving in the German city 8 days later. Commissioned by the Luftwaffe, during the War she was used as a replenishment and workshop vessel for its airfighters at Travemunde, Riga and in the fjord of Trondheim. After tak-

In the history of the Vlasov Group, the eventual *Castel Nevoso* had the privilege of introducing the name *Fairsky*, being the first of four vessels to bear this name; the picture shows her berthed at Purfleet, on the Thames, during Summer 1950.
(Laurence Dunn, Gravesend)

84

Castel Nevoso (1949-1968)

ing part in October 1940 in the air-naval operations off Brest and Bordeaux, she made her way back to Norwegian waters and was again based at Trondheim, then at Tromso and Bille. Here, on 19th September 1944, she was seriously damaged by a torpedo launched by a Russian fighter: her bow was completely blown up. The following November she was towed to the Narvik arsenal to be repaired, re-entering service on the following March at Trondheim. Captured by the British forces without being damaged, on 13th May 1946 the *Friesenland* was transferred to Wilhelmshaven and on 7th August 1947 she was commissioned by the R.A.F., arriving at Burntisland, in the Firth of Forth on 6th November.

In 1949 the *Friesenland* was bought by the Vlasov Group through its Alva S.S. Co. Ltd of London and on the last day of the year she left for the Deutsche Werft of Hamburg, where began her transformation into a refrigerated ship.

During her staying in the shipyard, on 24th February 1950 she was transferred to the ownership of the Alvion S.S. Co. and registered under the Panamanian flag with the name of *Fairsky*; she thus introduced the tradition of the forename "Fair" among the ships owned by Vlasov.

On 6th April the *Fairsky* left Hamburg heading for Amsterdam, where her fitting out was completed, with the addition also of ample and comfortable cabins for 24 passengers.

On 16th June 1950, with the refitting completed, she sailed for London, where she remained in lay-up for two months while her owner was discussing a long-term charter for her with the Italian Government for the import of bananas from

An aerial view of *Castel Nevoso* at speed, sporting on her funnel the logo of her charterers, Salen Rederei of Stockholm, although the shadow of the V is still visible.

Castel Nevoso (1949-1968)

Somaliland. At the time, indeed, a part of Somaliland was put under the trusteeship of Italy for a ten-year period by the U.N.O., guaranteeing in this way a continuity of privileged commercial relations between Italy and its ex-colony, from where 100% of bananas for the national needs was imported.

At the end of the conflict the Italian R.A.M.B., Regia Azienda Monopolio Banane (Royal Agency for the Monopoly of Bananas), although loosing the title of "Royal" in 1947, had continued to monopolise the import of this fruit. But, having lost all its modern banana-carriers during the War, the Italian agency signed charter agreements with a few private shipowners, among them Alexandre Vlasov, who had his *Fairsky* prepared for this purpose.

Arriving in Genoa on 23rd August 1951, three days later the *Fairsky* sailed on her maiden voyage to Chisimaiu, calling at Port Said and Aden. She continued on this regular route for almost seventeen years, completing 9-10 voyages per year, exception made for a few transatlantic voyages on charter in 1958 and between 1965 and 1968.

On 10th October 1952, to comply with the request of the Italian A.M.B., the *Fairsky* was transferred to the Italian flag and to Sitmar ownership with the new name of *Castel Nevoso*. This honoured the traditional forename 'Castle', given until then to all Sitmar's ships, while Nevoso ('Snowy') recalled the nature of the ship.

Furthermore, owing to the strong connection between Italy and its former colony, there was a constant number of businessmen and authorities travelling between the two countries and it was not difficult at all to fill the 12 double-berth cabins on board.

The bananas, practically the only goods loaded by the *Castel Nevoso*, were transported in jute sacks filled with straw to prevent damage. From 1958 onwards perforated cardboard boxes were introduced, which permitted a better protection and stowage of the fruit in the refrigerated holds. Upon every arrival in Genoa, the vessel discharged an average of 40-50 thousands bananas.

On 20th June 1958, on charter to Salen Rederei of Stockholm, the *Castel Nevoso* sailed on the first of two crossings to Santos. Back to Italy, on 18th September she re-started the African line to Chisimaiu, completing also four round trips to Tiko and Victoria.

She left her habitual route only at the end of 1965: on 22nd December, back to Naples from an occasional voyage to Falmouth and Greenock, she set sail on the first of seven consecutive trans-Atlantic voyages; one to Guayaquil, Ecuador via the Panama Canal, one to Buenos Aires and five to Canada. On 1st October 1966 she left Rotterdam for the United States, remaining in American waters until the following February and transporting fruit between Baltimore and New Orleans,

Castel Nevoso (1949-1968)

Note the modern flat transom stern of the Castel Nevoso *in this photograph taken in Genoa. (Victor Young & Len Sawyer, Wellington)*

sometimes pushing on as far as Golfito, Panama C.Z. On 24th May 1967 she was laid up in Genoa, at the end of a round voyage to Buenos Aires. Re-activated on the following 14th February she sailed for Bahia Blanca, with intermediate calls at Hamburg and Dublin. This proved to be her last voyage for Sitmar because, upon her arrival in Buenos Aires on 13th May 1968, she was laid up and put up for sale.

Although 30 years old she found a buyer who used her for further trading. In September she became the property of the Argentina Refeer S.S. Corp. of Panama and was re-named *Argentina Refeer*. On the 10th of the same month she started her new service along the East coast of South America, between Buenos Aires and Rio de Janeiro.

The new life of the former *Friesenland* as the *Argentina Refeer* was however a short one and on 5th May 1969 she left Buenos Aires for her last voyage, a crossing to the shipbreakers' yard of Faslane, where she arrived on the following 17th June.

CASTEL FELICE (1952-1970)

(formerly *Kenya, Hydra, Kenya, Keren, Fairstone, Kenya, Keren*)
Builders: Alexander Stephen & Sons Ltd (Glasgow)
12150 grt (7140 nrt) 5210 dwt; 1955: 12478 grt (7373 nrt) 6725 dwt;
151.80 [143.25] x 19.51 x 8.96 m (498.0 [470.0] x 64.3 x 29.4 ft)
2 sets of SR steam turbines; 9610 SHP; 17 kn
by builders
1952: 596 cabin passengers; 944 emigrants
1955: 28 cabin class, 1173 tourist passengers
1961: 1400 one-class passengers

The turbine steamer *Castel Felice*, which entered service for Sitmar in 1952, was the first real passenger ship bought by the Vlasov Group and she kept the honour as flagship of the fleet for almost 18 years, attaining a prominent place in the history and evolution of the company.

She was designed and built for the British India Steam Navigation Company at the end of the 'twenties, to consolidate their success on the India Africa run. The service had been maintained, from 1919 onwards, by the four liners of the "K" class: *Karoa, Karapara, Karagola* and *Khandalla*.

The new ship, the largest and fastest owned by the company up to that time, was laid down at the Linthouse yard at Govan in Glasgow, as no. 529 on the main building slip of Messrs. Alexander Stephen & Sons Ltd and was launched on 27th August 1930 as *Kenya*.

The *Castel Felice* leaving Genoa dressed overall at the start of her maiden voyage for Sitmar Line on 6th October 1952.

CASTEL FELICE (1952-1970)

Her hull had an overall length of 148.46 metres (487.1 ft) and, after the completion of the superstructure, she reached a gross tonnage of 9,830 tonnes.

A sister ship, the *Karanja*, was launched by the same yard on 18th December 1930, on the same day *Kenya* was running her delivery trials. On the measured mile she attained an average speed of 18.2 knots, against the 16 required by the contract. With a building price of £407,535 the *Kenya* was not only by far the largest and most modern ship of the fleet, but also, as far as the First Class passengers were concerned, the most luxurious, without falling into the trap of the hyper-decorativism of many other liners of her time.

Thanks to the Tropical and Equatorial routes for which she was conceived, *Kenya* was built with ample verandas and open-air promenades; the bright public rooms were equipped with large windows, cool upholstery and light panelling but no heavy drapery and fabrics.

The *Kenya* had cabin accommodation for only 66 First Class passengers and 180 in Second Class. Furthermore two decks were fitted with large dormitories and essential social rooms for about 1800 deck-passengers in order to satisfy the constant flow of Indians emigrating to East Africa.

The new flagship left Middlesbrough on 6th January 1931 on her delivery voyage to Bombay. She arrived at what was to become her habitual Indian terminus on 11th February, after calling at London, Antwerp, Southampton, Port Said, Suez and Aden.

The *Kenya* did not have a lucky start to her career as, while entering Durban for the first time, under the view of a big crowd waiting to see the new liner, she collided with the Union-Castle liner *Armadale Castle* suffering and causing slight damage; furthermore on 25th May, on her second voyage, en route from Zanzibar to Dar-es-Salam, she ran aground.

The *Castel Felice* started her life as the British India flagship *Kenya*, here seen departing Beira in September 1937.
(National Maritime Museum, Greenwich)

Castel Felice (1952-1970)

Then the service between India and Africa went on regularly with her sister ship *Karanja* until April 1940 when she was requisitioned by the British Ministry of War Transport and, the following month, she returned to Great Britain in convoy with a full load of Indian soldiers. She later took part in the unsuccessful Allied attempt of landing at Dakar and in the 5th May 1942 landings at Diego Suarez, Madagascar before being transformed into a "L.S.I.", Landing Ship-Infantry.

She was commissioned into the Royal Navy on 23rd July 1942 as H.M.S. *Hydra*, because a cruiser with her previous name was already in service. In October of the same year the ship was given a powerful transmitting apparatus to be used as Landing Ship Headquarters. Later her name was changed to H.M.S. *Keren*, after the Eritrean town where the Allied forces had recently won a battle against the Italians. The former *Kenya* emerged from this transformation with one six-inch and one three-inch gun, 22 anti-aircraft machine guns and 5 landing craft hunging on each side. She was also refitted to accommodate 1296 soldiers and a crew of 297 men.

In the following years she had an intense war career. In December 1942 she substituted for her sister ship *Karanja* (which had been bombed and sunk off Algeria on 12th November) in a series of convoy trips to North Africa. She later had an

During the Second World War the eventual *Castel Felice* won honours as the Royal Navy Infantry Landing Ship *Keren*.
(Imperial War Museum, London)

CASTEL FELICE (1952-1970)

important role as Headquarters Ship for the Bark East Sector in the invasion of Sicily, being one of the assault vessels in "Operation Husky", disembarking in July 1943 the 321st Infantry Brigade. She spent the following period of the conflict in Indian waters, taking part in the Burmese campaign and later in the Pacific, in the preparations for the invasion of Japan.

At the end of the War, British India considered replacing her with new tonnage and opted to accept the offer by the British Government to owners of requisitioned vessels to sell their ships to the Ministry of Transport. Thus the *Keren* was sold on 3rd April 1946 for £475,000, continuing in the repatriation of Allied Forces until August 1948, when she was laid up in Holy Loch, Scotland and put up for sale.

She remained idly at anchor until 19th February 1949 when, during a severe gale, she broke her moorings and was driven ashore. She was refloated and drydocked at her builders for rudder repairs, where Sitmar's representatives had a chance to inspect the ship and found her suitable to be converted for the emigrant service. In May she became the property of Alva Steam Ship Co. Ltd of London and was moved to a temporary anchorage in Rothesay Bay.

The acquisition of the former *Kenya*, a real passenger ship, designed and built for that purpose, was a good business move for Mr Vlasov. And thus, while the

The *Keren* off Pachino, Sicily, bearing the black stripe on her funnel, ensign of the Admiral's Ship, during the historic July 1942 Allied landings in Italy. *(Imperial War Museum, London)*

CASTEL FELICE (1952-1970)

British India was designing a new *Kenya*, a conversion study was started on the old one. A call was made for tenders from various yards. As the operation was very expensive things went slowly, giving the owner the chance to made up his mind that *Kenya*, by far the largest and best ship of the Group, could become something more than a vessel with only steerage dormitories.

Owing to many factors, such as new rebuilding plans and change of flag, the ship's name was exchanged between *Keren* and *Kenya* many times.

In 1950 she went back to Holy Loch and, with the changeover to Panamanian registry, she became the *Fairstone*, but only for a few months as in July she again took her first name and in October was registered in the maritime district of Rome for the Sitmar Line of Milan. In March 1951, with the name *Keren*, she left Holy Loch at last, in tow for Falmouth where the protective steel plates and the platforms for the guns added during her War service were removed and the promenade windows were opened again.

After being drydocked to inspect for seaworthiness on 10th March 1951 the liner was towed to Antwerp and on 22nd August she entered Genoa, where, at Cantieri Riuniti del Tirreno, definitive plans for the rebuilding were drawn up. Her power plant was made operational, new auxiliary equipment was installed, a new bow was fitted and some upperworks altered, including a new funnel.

The open-air promenade on the Saloon Deck of the *Castel Felice*. *(Jan Loeff, Fort Lauderdale)*

In September 1952 the old *Kenya* was reborn as the *Castel Felice* with 596 Cabin Class berths and accommodation for 944 Third Class passengers. On 6th October 1952 she set sail from Genoa for her maiden voyage to Australia and, after calls at Fremantle and Melbourne she arrived at Sydney on 7th November, a week before the scheduled arrival date.

On 9th December she was back at Genoa and was transferred to the Central and South America route but, starting on 13th July 1954, she was used also for two crossings from Bremen to Quebec as well as a series of voyages from Le Havre and Southampton to New York.

In May 1954, when the French fortress of Dien Bien Phu (Vietnam) was defeated by the Giap's warriors, the *Castel Felice* was sent to the Gulf of Tonkin to evacuate the French Foreign Legion.

CASTEL FELICE (1952-1970)

The *Castel Felice*'s Dining Room "A" in 1952; note the room is already fitted with a sprinkler fire-fighting system and that glasses and dishes bear the Vlasov "V". *(Jan Loeff, Fort Lauderdale)*

On 7th October 1954 she left Bremerhaven on a voyage to Melbourne, before being laid up in Genoa for further refurbishing of her interiors, the installation of an air-conditioning plant and of an open-air swimming pool aft. The *Castel Felice* re-entered service on 28th January 1955 with a crossing to Buenos Aires. On the return leg of the trip she steamed up the Adriatic to Trieste from where, on 26th February, she sailed for Sydney. She made two further trips to Australia in September and in March 1956 before being transferred to the New York service, from where she departed for the last time on 19th September 1957. In that

The *Castel Felice*'s Dining Room "B"; note the teak-cladded floor. *(Jan Loeff, Fort Lauderdale)*

Castel Felice (1952-1970)

The Captain's private dining room; a photograph of Alexandre Vlasov is hanging on the wall. (Jan Loeff, Fort Lauderdale)

As flagship of the Sitmar Line, the funnel of the Castel Felice was proudly kept immaculate by her crew.

year Sitmar signed an agreement with the British Government to transport migrants to Australia and, after a further revamping in Genoa, the *Castel Felice* started the Southampton-Sydney line service which she would successfully maintain for the next twelve years.

The *Castel Felice* was never intended for 'cruises' but, in December 1964 she made the first, unforeseen cruise in Sitmar's history. She was in Melbourne when a P&O liner, preparing to depart on a jaunt to Noumea, Cairns, Hayman Island and the East Coast was stricken by engine trouble; P&O organised what was probably the quickest charter agreement in maritime history and their passengers were promptly transferred to the Sitmar vessel. The Tourist Class accommodation, the service and the on-board cuisine were of a high standard which earned the ship a good reputation in Australia: on the return voyages to Europe the *Castel Felice* was usually full of young "Aussies". For this reason the itinerary could often vary, including a number of stop-overs in the Pacific islands as well as the transit of the Panama Canal.

On 15th August 1970, while the liner was berthed at Southampton, a fire broke out in the passenger accommodation. As the agreement for the transport of emigrants would have expired at the end of the year and the retirement of the old vessel had already been programmed, it was decided to continue the voyage but

CASTEL FELICE (1952-1970)

without repairs and with a reduced number of passengers on board.
On 26th September the *Castel Felice* reached Sydney, where her passengers disembarked and her furniture and stores were removed. On 7th October she left the Australian city bound for Kaoshiung where she arrived two weeks later to be broken up by Chou's Iron & Steel Co.

A 1960 view of *Castel Felice* in Singapore while embarking the harbour pilot.
(Ambrose Greenway, London)

FAIRSKY (1958-1977)

(formerly *Steel Artisan*, *USS Barnes*, *HMS Attacker*, *Castelforte*, *Castel Forte*,
then *Fair Sky*, *Fairsky*, *Philippine Tourist*, *Fair Sky*)
Builders: Western Pipe & Steel Company (San Francisco)
12464 grt (6682 nrt) 4950 dwt; 153.00 [141.79] x 21.18 x 7.77 m
(502.0 [465.2] x 69.5 x 25.5 ft)
2 sets of DR steam turbines; 8500 SHP; 17 kn
by General Electric (Lynn)
1461 one-class passengers; 248 crew

The second *Fairsky* owned by the Vlasov Group was, similarly to their first *Fairsea*, a conversion from an American C3 standard hull. However the appearance of the *Fairsky*, which entered service in 1958, was quite different from that of the *Fairsea*, which made her maiden voyage nine years before. Thanks to the better financial position of her owner and to the increased demands of the Tourist market she did not become an emergency converted emigrant ship like the *Fairsea* but a fine passenger liner. Her conversion was even more extensive than those carried out by the Naples shipowner Achille Lauro on his liners *Roma* and *Sydney*, two other C3 standard hulls, used as escort aircraft carriers during the Second World War.

The newbuilding destined to become the *Fairsky* was laid down on 7th April 1941 for the Isthmian Lines of New York as yard n. 62 on a San Francisco side-slipway of the Western Pipe & Steel Company (the builders of the ship which

The *Fairsky* with all her flags flying while leaving Auckland; note the huge name board on the bridge roof and the funnel exhaust extension.
(Victor Young & Len Sawyer, Wellington)

Fairsky (1958-1977)

eventually became Lauro's *Sydney*). She was launched on the following 27th September as *Steel Artisan*, one of the four C3 freighters intended for her owners. Instead of being delivered to Isthmian Lines as planned in February 1942, the *Steel Artisan* was transformed into the escort aircraft carrier U.S.S. *Barnes* and, on the following 30th September, she was loaned to the Royal Navy who commissioned her as H.M.S. *Attacker*.

After running her sea trials on 12th November 1942 she reached Liverpool on 4th April 1943 where further works were done to adapt the ship to Royal Navy standards.

In September she briefly took part in the Allied landings at Salerno (operation "Avalanche") but on 10th October she was back at Rosyth Dockyard to be transformed into an assault carrier. Only on 14th May 1944, after special training of her 646 crew and pilots, she took to sea again, participating in operation "Dragon", the invasion of the South of France. The H.M.S. *Attacker* was later involved in the strikes against German forces in the Dodecanese Islands and the Aegean Sea (operations "Outing", "Cablegram" and "Contempt") and on 13th

The *Fairsky* was re-built from the hull of the C3 auxiliary aircraft carrier U.S.S. *Barnes*; the picture depicts her after having been converted in 1944 by the Royal Navy into the assault-carrier H.M.S. *Attacker*.
(Imperial War Museum, London)

Fairsky (1958-1977)

The *Castel Forte* in March 1958 at the Varco Chiappella fitting out berth in Genoa, undergoing her transformation into the *Fairsky*; close to the stern gangway and on the sides of the forecastle are well visible her name boards.
(Giorgio Ghiglione, Genoa)

October 1944 was one of the assault carriers in the operation "Manna" for the re-occupation of Athens. After a period of five months spent at Taranto undergoing repairs and a refit, she left the Italian port on 1st April 1944 bound for the Pacific. Here she took part in her last War mission, operation "Juris", for the re-occupation of Singapore. This successfully completed, on 11th November 1945 she arrived on the Clyde. Fully returned to the U.S.A., she arrived at Norfolk on

Fairsky (1958-1977)

the following 5th January and was laid up. On 28th October 1948 she found a buyer in the National Bulk Carriers Co. of New York but the plan to transform her into a freighter did not materialise and in 1950 she was re-sold to the Vlasov Group through their American Navcot Corporation which re-named her *Castelforte* and registered her under the Panamanian flag.

As was common with all the other merchant vessels of the time, to authorise the sale to a foreign flag the American Maritime Commission asked for the conversion works to be carried out in a national shipyard. Thus in 1952 the ship entered the Newport News Shipbuilding and Drydock Co. where a plan was drawn up for her conversion into a refeer ship for the transport of meat from Argentina to Europe. The works were however soon stopped and the *Castelforte* started a long period of lay-up during which nothing more was done except such maintenance as was necessary to keep the ship alive.

It was not until February 1957 that the yard's workers again boarded the vessel to transform her into a passenger ship, whose name had been amended to *Castel Forte* in 1954. The reason why Mr Alexandre Vlasov decided at last to put her into service stemmed above all from the sale at that time of the *Castel Verde* and

The *Fairsky*'s main lounge designed by Mr Störmer from Bremen; for the first time in its history, Sitmar Line engaged a professional architect to design the ship's interiors.

The *Fairsky* steaming out-bound Singapore in 1960.
(Ambrose Greenway, London)

Fairsky (1958-1977)

The swimming pool of the Fairsky.

Castel Bianco to the Compañía Trasatlántica Española.

On 26th January 1958 the *Castel Forte* arrived in Genoa under the command of Capt. Jorge Petrescu with a crew of 43 men and berthed at the Calata Chiappella fitting out quay. The hull had already been modified by the insertion of a new raked stem at Newport News, where the engine plant had also been overhauled and the new auxiliary equipment (including the air conditioning compressors) had been installed on board. At the time the upperworks consisted only of a small central house surmounted by the bridge and a cylindrical funnel.

To the Genoa-based T. Mariotti works was entrusted the co-ordination of the 120 Italian firms involved in her transformation into a passenger liner; they would give the *Castel Forte* a distinctive, modern profile, influenced by the beautiful Ansaldo-built passenger ships of her time. The upperworks consisted of three new decks terminating forward with a front shaped like that of the *Cristoforo Colombo* and the *Federico C.* which was being completed at the same time at a nearby fitting out quay. Also the mast over the wheelhouse and the funnel casing were reminiscent of recent Ansaldo ships, as well as the distribution of the accommodation on board.

The air conditioning was extended to the whole vessel, including the crew quarters, and the most up-dated fire prevention and fire fighting systems were installed. The interiors provided accommodation for 1461 one-class passengers and 250 crew members; there were in all 461 cabins (189 outside) with six, four or two berths each but only seven de-luxe ones, on the Sun Deck, had private facilities.

The three passenger dining rooms with their galley occupied the entire Promenade Deck, while on the upper Boat Deck there were the Lido with open-air swimming pool and all the public lounges, exception made for the Entrance Hall (Deck A) and the Veranda Bar (Sun Deck). A beautiful sheltered promenade with teak flooring ran around the three bright dining rooms and similarly, on the upper Boat Deck, another large promenade (acting as muster station in case of emergency) was around the public lounges.

Between 6th and 13th May 1958 the *Castel Forte* was drydocked in Genoa and it was at this point that the boards with the old name were removed and the new one, *Fairsky,* was painted on her hull. At the same time she was transferred to the Sitmar subsidiary Fairline Shipping Corp. of Monrovia, keeping however the Panamanian flag.

Fairsky (1958-1977)

At 14.20hrs of 19th June 1958, again under the command of Capt. Petrescu and with 206 crew members on board, the *Fairsky* left Genoa for Southampton where she arrived 5 days later. There she embarked her first passengers and on the following 26th she left on her maiden voyage via the Suez Canal to Brisbane, under charter to the British Government for the transport of emigrants to Australia, calling at Le Havre, Fremantle, Melbourne and Sydney, where she arrived on 29th July.

During Easter 1965 the *Fairsky* departed on her first cruise: a two-week jaunt from Sydney to Noumea, Cairns, Hayman Island and Brisbane.

The vessel remained in regular Australian service for 14 years although, after the Six Day War of June 1967 between Israel and Egypt, the Suez Canal was closed to merchant traffic and the steamer was re-routed via Cape Town outbound and via the Panama Canal on the return leg.

In August 1968 her name was slightly modified to *Fair Sky* and, in November 1969, while approaching Cape Town, she suffered engine troubles and was forced to make a 2-week stop in the African harbour to undergo repairs.

Although the Vlasov contract with Great Britain for assisted passages was lost in 1970 to the Greek Chandris Line, the *Fair Sky* maintained the regular service to Australia until February 1972, when she was laid up at Southampton. Her owners had at the time ready for operation two fine new passenger ships, the second

On 23rd June 1977 the *Fairsky* was deliberately stranded by her master to prevent her sinking after fouling the remains of a sunken vessel.
(Roberto Giorgi, Monaco)

Fairsky (1958-1977)

Fairsea and the *Fairwind* (ex-Cunarders *Carinthia* and *Sylvania*) and so the future of the *Fair Sky* seemed uncertain. Eventually it was decided not to put the new twins on the Australian line as initially planned but to use them instead to enter the then booming and lucrative cruise market in Florida.

The liner, again named *Fairsky*, could thus re-enter service on her habitual route on 8th November 1973, with a regular voyage Southampton-Sydney. However, after a few crossings, in July 1974 she was finally based at Sydney as a full-time cruise ship: she had a great success, opening the way to her fleetmate and successor *Fairstar*.

Unfortunately her career with Sitmar had an abrupt and unforeseen end. On 12th June 1977 the *Fairsky* left Darwin for what was to became her last cruise: on 23rd June, while leaving the Djakarta harbour of Tandjung Priok, her keel fouled the submerged remains of the Indonesian combi-ship *Klingi* and a big breach was opened up in her bottom. To avoid the danger of sinking the ship was deliberately beached by her master and on the 29th she was refloated. Under her own power she reached a Singapore drydock but after inspecting her, the shipowners decided that it was not worthwhile carrying out definitive repairs: a temporary patch was fitted over the hole and she was laid up and put up for sale.

She was eventually sold to the Hong Kong shipbreakers Fuji Marden & Co. and on 11th December 1977 she left Singapore on what seemed to be her last voyage. But before demolition started, she was bought by the Filipino firm of Peninsula Tourist Shipping Corp. which re-named her *Philippine Tourist* and transferred her to the Bataan Shipyard & Engineering Co. of Manila to be transformed into a permanent floating hotel and casino. Unfortunately her new career was all too short as, on 3rd November 1979, a fire broke out and destroyed the upperworks and the interior accommodation of the former *Fairsky*. In a curious twist of fate she was re-named *Fair Sky* and towed back to Hong Kong to be broken up by the same Fuji Marden & Co. from which she had escaped two years before.

The last active duty of the former *Fairsky* was a very short stint as the hotel and casino ship *Philippine Tourist*, permanently moored in Manila.
(Roberto Giorgi, Monaco)

FAIRSTAR (1964-1988)

(formerly *Oxfordshire*, then *Ripa*)
Builders: Fairfield S.B. & Eng. Co. Ltd (Glasgow)
21619 grt (12480 nrt) 8800 dwt; 185.72 [170.68] x 23.77 x 8.41
(609.3 [560.0] x 78.0 x 27.6 ft)
2 Pametrada geared turbines sets; 18000 SHP; 18.5 kn
by builders
1868 one-class passengers; 460 crew
1984: 1390 cruise passengers
1989: 1280 cruise passengers

On 12th February 1997, the *Fairstar*, Australia's most famous and most beloved cruise ship, with the name *Ripa* (Rest In Peace Always) painted on her hull and flying the St. Vincent & Grenadines flag, cast off all lines from the Darling Harbour terminal, passed the Harbour Bridge and the Opera House and slowly steamed down Sydney Bay for the last time, en route to an Indian shipbreaker's yard.

It was a sad and historic date, for the *Fairstar* was for more than three decades the Australian cruise ship par excellence and a very familiar sight to all Sydney's inhabitants, like one of the many buildings which embellish that beautiful harbour. With the *Fairstar* has gone the last classic liner of the Sitmar Line, which for a decade served the regular passenger route between Europe and Australia and

The *Fairstar* in Cape Town harbour during her liner service to Australia.
(Matteo Parodi, Monaco)

Fairstar (1964-1988)

On 15th December 1955, the Oxfordshire left the Govan slipway of the famous Fairfield shipyard. (Glasgow City Council Archives)

later successfully introduced Sitmar to the cruise market. The *Fairstar*, under the command of Capt. Jorge Petrescu, entered Sydney Harbour for the first time on 7th June 1964 on her maiden crossing from Southampton.

She was at the time under charter to the British Government for the transport of migrants, almost all of whom disembarked during the calls at Fremantle, Adelaide and Melbourne, from where the new Australian citizens were distributed round the whole territory.

But before becoming Sitmar's *Fairstar*, the vessel had a previous life as the troop transport *Oxfordshire*.

She was originally designed in the early 'fifties with a similar vessel, the *Nevasa*, to partner H.M. Troop Transports *Dilwara*, *Dunera* and *Devonshire*. At the time the United Kingdom was still an important imperial power and the traffic of soldiers to and from its colonies was of essential importance to the survival of the Empire. In the inter-War period the British Ministry of Transport had entrusted P&O, the British India Steam Navigation Co. and Bibby Line with the management of peacetime trooping. The latter company had a very long tradition of troop transports, started during the Crimean War of 1854. In 1951, when a financial offer by the Government made a new vessel attractive to them, the 1912-built *Oxfordshire* was sold and a new one ordered from the Fairfield Shipbuilding and Engineering Co. of Glasgow.

The new *Oxfordshire* was laid down as hull 775 at the builder's Govan yard and launched on 15th December 1955 by Mrs Dorothea Head, wife of the Minister of Defence.

Fairstar (1964-1988)

The *Oxfordshire*, designed for a service speed of 17 knots, ran her delivery trials on 29th January 1957, reaching an average speed of 20.93 knots on the measured mile, corresponding to a power output of 19,875 SHP at 121.5 rpm of the two propellers.

Delivered to her owners on 13th February 1957, she could accommodate 1000 troops and 500 passengers (the soldiers' families) in three classes plus a full complement of 409 crew members. Her near-sister *Nevasa*, had been delivered to British India the previous July: they were the largest (and last) purposely-built troop transports ever launched for Great Britain.

On 28th February the *Oxfordshire*, under the command of Capt. N. F. Fitch, left her home port of Liverpool on her maiden voyage to Hong Kong via Cape Town, calling at Dakar, Durban and Singapore.

The *Oxfordshire* and the *Nevasa* were originally intended for a 15-year charter to the Ministry of Defence but actually, owing to the rapidly changing political situation and the dissolution of the Empire, they would ply the routes between their mother country and its colonies for five years only. In October 1962 the *Nevasa* was withdrawn from service and the *Oxfordshire* closed this page of history two months later when she disembarked at Southampton 520 soldiers of the Royal Highland Fusiliers and their 156 relatives, who had boarded the ship at Valletta, Malta, which was soon to become independent. With the Australian passenger

The *Fairstar* started her life as Bibby Line's *Oxfordshire*, the last British troop tranport to be built. (Victor Young & Len Sawyer, Wellington)

Fairstar (1964-1988)

Fairstar (1964-1988)

trade then booming and the contract from the British Government renewed, Sitmar showed an interest in the *Oxfordshire* and in February 1963 a six-year charter agreement with a final option to acquire the liner was signed between the Bibby Line and the Vlasov Group's subsidiary Fairline Shipping Corporation of Monrovia.

On 17th May 1963 the *Oxfordshire* left Falmouth, where she was in lay-up, and two days later entered the Wilton-Fijenoord yard at Schiedam.

The plans to convert the *Oxfordshire* into a double purpose vessel, suitable for both liner voyages and cruises, was one of the most radical ever prepared by the Vlasov technical office, the first one drawn up in the new headquarters in the Principality of Monaco and also the first after the death of Alexandre Vlasov. She was, in some ways, the first ship created by Boris Vlasov; he entrusted the financial management of his empire to Ardavast Postoyan (employed with Vlasov since November 1938) and he personally supervised the transformation, known at the time as "Conox Project" (Conversion of *Oxfordshire*).

Boris Vlasov had in his mind a very high standard for the *Fairstar*, to cater for the highly demanding paying passengers and to face the competition posed by luxury liners on the Australian market, in particular the Lloyd Triestino's sistership *Galileo Galilei* and *Guglielmo Marconi*.

The conversion proved to be longer, more difficult and more expensive (£4.5 million) than forecast and to speed things up it was first decided to buy out the vessel in March 1964 (solving the problem of crew rosters raised with the Bibby Line) and to move her to Southampton. Here Harland & Wolff would finish the fitting out, previously delayed by the extreme weather conditions in Holland and also by a fire which broke out on 29th December 1963.

When the former *Oxfordshire* was delivered to her new owners in May 1964 she had a new and different look. The superstructure was extended fore and aft, with the elimination of the 3 pairs of cargo-booms, substituted by two cranes on the forecastle. The traditional mast was also replaced by a streamlined one of modern design and a central house was added around the funnel, giving to it a more squat appearance.

The most radical transformation was, however, inside and was designed by architect Störmer from Bremen. The troop dormitories made way for 484 cabins, 420 with private facilities, distributed on six decks. Most of them had four berths but on the Boat Deck there were also four luxury suites, with separate sitting and bed rooms, named after famous constellations.

The ship was now fully air-conditioned; in fact, according to Sir Derek James Bibby, when the design of the new troopship *Oxfordshire* was being discussed with the Ministry of Transport, they insisted that no air conditioning should be

Previous page:
a group of yard workers observes the arrival of the just waterbone *Oxfordshire* at the fitting out quay, carefully towed by a tug through the mist of a drizzling Scottish Winter.
(Glasgow City Council Archives)

A spartan troop dormitory on the *Oxfordshire*; note the foldable triple beds and the metal shelves for the soldiers' knapsacks.
(Glasgow City Council Archives)

The Brisbane Cairncross drydock projectors illuminate the *Fairstar* during her June 1984 three-week refit, during which a blue stripe was also added to her white hull.
(Mario Bertolotti, Monaco)

installed, even though many families travelled on board. This was to ensure that troops were tropicalised by the time they reached the Far East.

The *Fairstar* could accommodate 1868 one-class passengers who had the use of many well-furnished public lounges, mainly fitted on the Promenade Deck and flanked by sheltered teak-clad side promenades. The main lounge, called the Zodiac Room, spanned two decks linked by two sweeping staircases. Among the other amenities on board there were the Aquarius Bar, a night club fitted with portholes looking into the swimming pool, and a large cinema-theatre with 367 seats. But possibly the most liked rendezvous by her Australian passengers was the Tavern, a German-style beer hall, better known as Animal Bar; during her last cruise many *Fairstar* enthusiasts proudly wore the most popular T-shirt ever sold on board displaying the inscription "I survived the Animal Bar of the *Fairstar*".

On 19th May 1964 the *Fairstar* left Southampton on her maiden voyage to Sydney. She maintained this regular liner service for almost nine years, although, after the Six Days War of June 1967 the Suez Canal was closed and all the passenger liners on the Australian line, including the *Fairstar*, were re-routed via Cape Town eastbound and via the Panama Canal westbound.

From her first year with Sitmar the liner was used in low-season for cruises: the first one was under charter to the Massey-Ferguson firm for their annual convention of January 1965. The following December she made a successful 11-day

Fairstar (1964-1988)

Christmas and New Year jaunt to Noumea and Fiji, immediately followed by another 3-week cruise to Tahiti.

With the passing of time the *Fairstar* was used more and more for cruises and in July 1973 she sailed from Southampton on her last liner voyage. The following month she was based at Sydney as a full-time cruise ship, sailing on the 24th under the command of Capt. Rodolfo Potenzoni for a 16-day jaunt to Savu Savu, Lautoka, Suva, Auckland and Melbourne with a full complement of 1356 passengers. In May 1974, Sitmar, in its search for new market niches, despatched the *Fairstar* to Southampton for an experimental season of cruises from the English port. However it proved much more difficult to fill the vessel there and on 13th November she left European waters for ever and returned to Sydney.

Between 3rd and 21st June 1984, after the entry into service of the new *Fairsky*, the *Fairstar* was in the Cairncross drydock in Brisbane to be re-styled and upgraded, in order to match the new flagship's higher standard of accommodation. In the process the total number of passengers was reduced to 1390 people and, curiously enough, to extract the fin stabilisers and inspect them, two large niches were opened in the drydock side walls.

In November 1984, Sitmar, after the success of a similar event attended by both *Fairsea* and *Fairwind* in 1977, organised a special cruise off Noumea to witness a total eclipse of the sun, which was strongly advertised in Australian newspapers. For the occasion some American scientists were on board with their powerful telescopes. Unfortunately, a boiler failure which occured during the last day at sea of the previous cruise, delayed the start of the voyage from Sydney and when the *Fairstar* reached the position the extraordinary event was missed, to the consternation of all her passengers. On the day the vessel arrived back in Sydney a rival cruise company bought a page on the local newspaper to publish in capital letters "*Fairstar* missed total eclipse". After this accident the vessel's boilers were re-tubed at the Sembawang shipyard at Singapore.

As a proof of the extraordinary popularity of the

A standard cabin and a dining room of the *Oxfordshire*, which remained almost unaltered on the *Fairstar* during her early days as a Sitmar liner.
(Glasgow City Council Archives)

The *Fairstar* tendering passengers off Mystery Island, Vanuatu, on 16th April 1991; the white swan on her blue funnel, which replaced the V in early June 1988, was changed to a blue swan on white background in mid-June 1991.
(Alberto Bisagno, Genoa)

The much awaited ceremony for the crossing of the line, as seen from the throne of King Neptune.
(Alberto Bisagno, Genoa)

Fairstar (1964-1988)

Fairstar, she was the only Sitmar vessel to retain her name after the 1988 acquisition by P&O, which maintained for the vessel a separate administrative branch office in Sydney and used a special funnel logo. The first order of business planned for the *Fairstar* by her new owners was an extensive refit.

On 19th April 1989 she entered the Sembawang shipyard of Singapore, where she had her usual biannual drydocking and overhaul, emerging with an extended-aft Boat Deck, the total revamping of public lounges and cabins and a new potable water plant. In the process, the total number of berths was again reduced, this time to 1280.

On 19th June 1991, while steaming off the Vietnamese coast on a 29-day cruise to Hong Kong, Korea and Japan, her boilers failed owing to an infiltration of salt water. The vessel remained adrift for two days in the South China Sea and, with the air-conditioning not working, she became like a Turkish bath. The passengers had to camp on the open decks until the Australian deep-sea tug *Lady Sonia* ferried them to the coast, while the *Fairstar* was towed to the Singapore shipyard which, curiously enough, she had left a few days before for an engine overhaul. This unfortunate episode compelled P&O to refund the 1150 passengers and to cancel three cruises.

On 31st January 1997, the *Fairstar* concluded her last cruise in Sydney; a long white farewell pennant flies from her mainmast while in the background her old fleetmate *Fair Princess* is ready to take her place.
(Gerald Laver, Leongatha)

Fairstar (1964-1988)

On 12th February 1997, the Fairstar *left Sydney bound for the shipbreaker's yard; the dolphin on the funnel and her name have been painted over, while* Ripa *was quickly hand-painted on her hull and upperworks' name boards.*
(Gerald Laver, Leongatha)

In June 1994, *Fairstar* had the honour to inaugurate the new Sydney passenger terminal at Darling Harbour, in replacement for the old Pyrmont maritime station. The new terminal has the privilege of being only a few minutes walk from the main town's tourist attractions, such as the aquarium, the maritime museum, the shopping centre and the mono-rail panoramic train station.

In later years the *Fairstar* suffered frequent engine breakdowns and the average speed during cruises was greatly reduced to avoid a total collapse of her old machinery. In April 1996 the *Fairstar* failed to sail on a programmed cruise owing to a sudden fall of boiler brickwork; inspection showed that extensive and costly repairs were needed, also to bring the vessel up to 1997 Solas requirements, and soon afterwards P&O announced its decision to withdraw the vessel in early 1997. In a couple of days her last 10-day cruise, departing Sydney on 21st January, was sold out.

FAIRSEA (1971-1988)

(formerly *Carinthia*, *Fairland*, then *Fair Princess*)
Builders: John Brown & Co. Ltd (Glasgow)
21916 grt (12228 nrt) 6742 dwt; 185.40 [173.74] x 24.40 x 8.94 m
(608.3 [570.0] x 80.1 x 29.3 ft)
2 sets of Pametrada DR steam turbines; 24500 SHP; 20 kn
by builders
1971: 884 passengers, 470 crew

In the second half of the 'sixties Sitmar had to face the replacement of two vessels in its fleet, the *Castel Felice* and the *Fairsea*: they had been born as emergency emigrant vessels after the War and now they could no longer compete with the liners of rival companies or with their own fleet-mates *Fairsky* and *Fairstar*.
The occasion to replace the two old vessels presented itself at the end of 1967, when the Cunard Steam Ship Co. Ltd closed down its regular line between Great Britain and Canada, placing on the sale list the *Carinthia* and the *Sylvania*.

The 1971 *Fairsea* was the first passenger ship which never operated any liner voyage for Sitmar.
(William H. Miller, Secaucus)

Fairsea (1971-1988)

The Carinthia *being fitted out at a busy John Brown's shipyard.*
(National Maritime Museum, Greenwich)

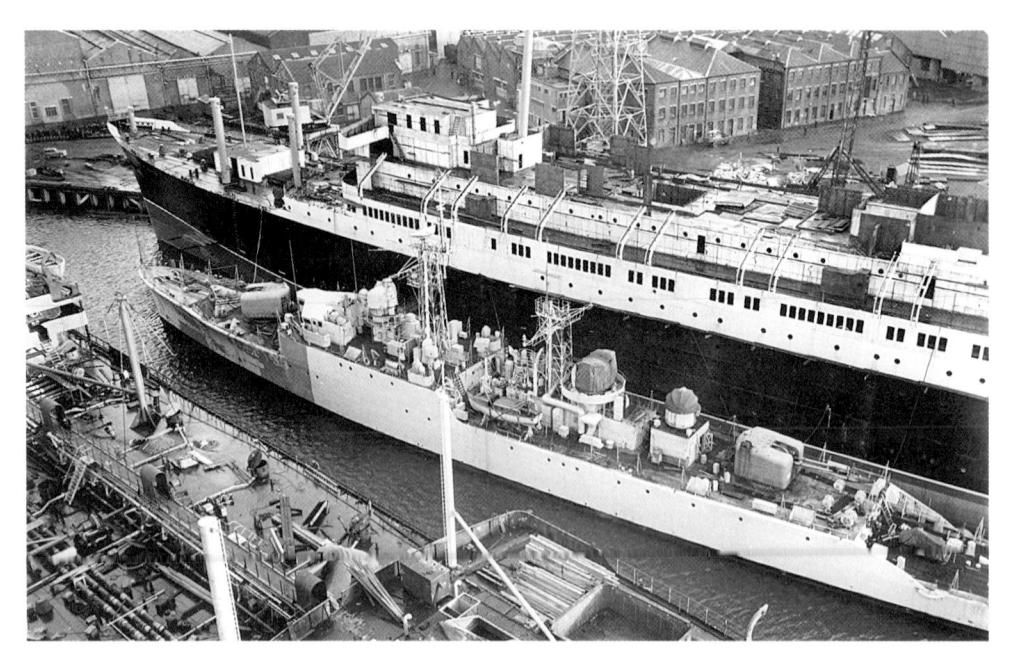

Bought by Sitmar, they would become the second *Fairsea* and the *Fairwind*, the first ships in the company history exclusively intended for cruises and which never operated any liner voyages.

Carinthia and *Sylvania* were two of four sisterships built by John Brown & Co. Ltd between 1954 and 1957 for the Cunard Line's passenger and freight express service to Canada.

The most important route operated by the Cunarders after the New York one, the regular line between Liverpool and Montreal was re-opened after the War by the

The Fairsea *entered service in 1956 as the Cunard's* Carinthia *to operate on their regular Liverpool-Montreal line.*
(Alex Duncan, Gravesend)

Fairsea (1971-1988)

venerable *Aquitania*, eventually sold for scrap at the end of 1949. However, in June 1949 the 26-year old *Franconia* joined her after an expensive refurbishing and within two years also the pre-War *Ascania*, *Scythia* and *Samaria* were all in regular Canadian service.

These old steamers were however intended as stop-gap passenger ships, and soon Cunard decided to replace all of them with new tonnage. In December 1951 they ordered the sisterships *Saxonia* and *Ivernia*, with an option for a further pair of twins, the eventual *Carinthia* and *Sylvania*, from the famous Clydebank shipbuilders. In all, they were sound vessels of traditional design, both in their external appearance and in the distribution of their interior spaces. They had a well balanced, fine profile, although rather conservative compared with other contemporary ships, already designed for a possible use in the cruise market. They could accommodate 154 passengers in First Class and 714 in Tourist, plus a crew of 461. The power plant was based on the traditional double set of double-reduc-

The *Fairwind* (left) and *Fairland* (later re-christened *Fairsea*) remained in lay-up in Southampton for a long time at the end of the 'sixties, before being transformed into two of the most successful cruise ships ever, now in service for almost thirty years.
(National Maritime Museum, Greenwich)

FAIRSEA (1971-1988)

Encircled by scaffolding, the minium-red painted *Fairsea* awaits her final colours in Trieste. *(Arthur Crook, Forest Row)*

tion steam geared turbines developing 24,500 SHP and a speed of 22 knots.

The ship which eventually became Sitmar's *Fairsea* was the only one of Cunard's famous "quads" to have a royal launching. Laid down as yard no. 699, she was named the *Carinthia* and sent down the slipway by HRH Princess Margaret on 14th December 1955.

Although virtually identical outside, the four sisters differed in their interior décor; the *Carinthia* (and later the *Sylvania*) were fitted with a higher standard of accommodation than that found on board the first two vessels, *Saxonia* and *Ivernia*. They remained ships with a traditional atmosphere, however, and actually part of the furnishing on board the *Carinthia* (including the chairs of the First Class dining room) came from the *Aquitania*.

On 27th June 1956, under the command of Captain Andrew McKellar, the *Carinthia* sailed on her maiden voyage to Montreal. In winter months, when the St. Lawrence river was impracticable, she was re-routed to Halifax and New York, from where she sailed in December 1956 for a Christmas and New Year cruise to Caribbean waters, the first of the sisters to be used for pleasure jaunts.

In the early 'sixties, competition from the jet aircraft developed to such an extent that Cunard were making a continuous loss on the operation of their liners: in

Fairsea (1971-1988)

An interesting aerial view of *Fairsea* during her maiden arrival in Los Angeles. *(William H. Miller, Secaucus)*

low-season the *Carinthia* and her three sisters often had only a few dozen passengers on board and a cargo of a few hundred tons against the 10,000 that the ten holds could store. In particular 1961 was not a lucky season for the *Carinthia*: in January, while laid-up in Liverpool for her biennial 2-week overhaul, a strike by the ship-repair workers blocked her in the drydock for four months. On the following 30th August, while steaming up the St. Lawrence in thick fog, she collided with the Canadian freighter *Tadoussac*, reporting damage to her hull and superstructure.

At the end of 1962 the *Saxonia* and the *Ivernia*, which had usually run out of

117

Fairsea (1971-1988)

The spacious and comfortable dining room and one of the ample double-berth outside standard cabins on board the Fairsea. (William H. Miller, Secaucus)

Southampton rather then Liverpool, were withdrawn from the liner service to be refitted as cruise ships. The *Sylvania* had been transferred to the New York run the previous April: thus the *Carinthia* remained the last trans-Atlantic liner working on the Liverpool-Montreal route.

For the Merseyside city, with its economy based for a century on its port, it was the definitive and bitter omen that a prosperous era was close to its end.

On 8th December 1966 the *Carinthia* was caught in the worst gale of her career. Arriving at Liverpool 48 hours behind schedule, she had to be drydocked for repairs, the most severe damage being to her rudder. After drydocking the *Carinthia* was scheduled to sail for the traditional Christmas and New Year cruise and for the occasion Cunard proposed to re-paint her in a white livery. The delay caused by the storm and the subsequent repairs prevented the execution of this work and the liner kept her black hull until the end.

On the evening of 13th October 1967 *Carinthia* left Liverpool for the last regular Cunard voyage to Canada, sealing a page of maritime history of tremendous meaning.

On 9th December, upon her return to her mother country, the *Carinthia* was laid up at Southampton's 101 berth and put up for sale. That year her owners had in fact lost 2 million pounds and Cunard's chairman, Sir Basil Smallpiece, announced that the company was prepared to sell the *Carinthia* and the *Sylvania* "without any strings attached to the sale" as soon as possible.

On 2nd February 1968 the two ships were actually bought by Sitmar Line for one million pounds each and re-named *Fairland* and *Fairwind*, respectively.

At the time the British contract for assisted passages to Australia was in the hands of Sitmar, but it would lapse in December 1969. For this reason the acquisition of the two Cunarders to replace the old *Fairsea* and the *Castel Felice* seemed a right move in order to obtain the 5-year renewal of the contract.

However, the decline of the traditional passenger service by sea in favour of airlines brought private

Fairsea (1971-1988)

shipowners to a quick rethinking of their marketing strategy for the future; in a few years they would all abandon liner service to dedicate themselves completely to the cruise market. Furthermore, at the end of the 'sixties the contribution by the British Government to the assisted passages out-bound to Australia covered only the ships' running costs, made irksome by the dizzy increase in the fuel price and by the increasingly scanty numbers of paying passengers northbound.

And thus the loss by Sitmar in 1970 of the migrant contract to the Chandris Line did not in the end affect the Vlasov Group negatively; on the contrary it stimulated the brilliant transformation of the Sitmar Line into Sitmar Cruises. At the time there were great changes in Sitmar's policy: the emigrant liners *Fairsea* and *Castel Felice* were sold for scrap, *Fairsky* and *Fairstar* quickly abandoned the Europe-Australia line and were permanently based in Sydney as full-time cruise ships, while at the new-building office of the Monte Carlo headquarters Boris Vlasov ordered the start of the "Concarsyl" project, "Conversion of *Carinthia* and *Sylvania*" into de-luxe cruise ships.

After a call for tenders from various European yards, the Arsenale Triestino San Marco shipyard was awarded the contract for the transformation of both the former Cunarders.

The *Fairland* arrived in Trieste on 21st February 1970 and two months later her

The *Fairsea* steaming into Willemstad, Curaçao.
(William H. Miller, Secaucus)

Fairsea (1971-1988)

The former Fairsea, now under the Princess banner as their Fair Princess, during a Vancouver cruise. (Victor Young & Len Sawyer, Wellington)

name was changed to *Fairsea*. Here the yard workers stripped out all the furniture, panelling, wooden partitions, floors, ceilings, insulation, as well as all the electrical cables and piping, leaving the vessel in a bare-steel condition. Outside, the distinctive Clydebank domed funnel was removed and all the hull sand-blasted to the metal, while 25-tons of hull plates damaged by bumps were replaced. All the upperworks, from the main deck up, were extended fore and aft, the cargo booms eliminated, a new streamlined funnel casing and a new mast installed.

All the new interior accommodation was planned to a grand scale and a very high standard under the supervision of the famous Italian architect Umberto Nordio. From the early stages of design particular attention was paid to safety. Thanks to a special agreement between Vlasov and Lloyd's Register, the latter's senior surveyor, Arthur W. Crook, in charge of the Passenger Safety Certification Department, was sent to Monte Carlo to work side by side with Sitmar's naval architect Dario Rivera and later to superintend the rebuilding at Arsenale

Fairsea (1971-1988)

Triestino. Their co-operation would bring *Fairsea* and *Fairwind* to comply fully with 1960 Solas rules; at the time their conversion was considered to be very important to both Lloyd's Register and the United States Coast Guard and in fact the Society sent a photographer on board to take many pictures for inclusion in their prestigious "100A1" magazine.

On 3rd November 1971 the *Fairsea* was delivered to a Sitmar subsidiary, Fairline Shipping Corp. of Monrovia, and on the 9th she embarked in Cadiz many tour operators for a trans-Atlantic cruise to Antigua, St. Thomas, Acapulco, Los Angeles and San Francisco, where she arrived on 10th December. Four days later, after being officially presented to the press, she made her first cruise to the Mexican Riviera. On 6th August 1972 she cleared Los Angeles for her first trans-Panama Canal cruise to the West Indies, while from June 1974 she spent the following summer seasons in Alaskan cruises, leaving from San Francisco.

In order to make their way in the competitive American cruise market, Sitmar Cruises made an enormous advertising effort, offering also an exceptional on-board service for both entertainment and cuisine compared with the prices they offered. Results soon came, and both *Fairsea* and *Fairwind* became well-known vessels beloved by the American market.

The *Fairsea* consolidated her success for a quarter of a century on three different itineraries: two from Los Angeles with destination the Caribbean via Panama Canal in Winter and the Mexican Riviera in Summer, and Alaska, again in Summer, using San Francisco as terminal port.

In Spring 1984, after the entry into service of the new *Fairsky*, *Fairsea* and *Fairwind* were both refurbished in the Norshipco shipyard of San Francisco to match the higher standard of the new flagship. The project was entrusted to the well-known Genoese architect Giacomo Erasmo Mortola, nowadays interior designer of the 80,000- and 100,000-ton giants built by Fincantieri for Princess Cruises. In 1981 he had founded in Genoa a company called Navstaff specifically to design some of the interiors of the *Fairsky*.

In 1988, with the sale of Sitmar Cruises, *Fairsea* became *Fair Princess*; upon completion of her scheduled cruises to Alaska, in September 1988 the liner was drydocked in Los Angeles for a partial refurbishing of her interiors. She then departed wearing the Princess Cruises livery on a special positioning cruise to Tahiti, New Zealand and Australia. On 10th November she berthed in Sydney for the first time, starting a series of weekly cruises which lasted until April 1989, when she made her way back to the United States, her subsequent itineraries remaining the same as those she had followed for Sitmar.

In 1995 the imminent sale of the ship to Regency Cruises to become their *Regent Isle* was announced. However, owing to the financial troubles of the latter com-

Fairsea (1971-1988)

In late 1996, preparing to substitute for the Fairstar *in Australia, the* Fair Princess *was refitted in San Diego to meet the new Solas requirements; her Princess logo appears over-painted in white and soon her funnel will receive a coat of P&O plain buff. (Victor Young & Len Sawyer, Wellington)*

pany, the sale never materialised and in 1996 the *Fair Princess* was chosen to replace the *Fairstar* in Australian waters.

In December 1996, after a period of lay-up, a six-month refit was completed at San Diego to meet the 1997 Solas rules, while the funnel was repainted in the plain buff colour of P&O. The *Fair Princess* made her first cruise from Sydney on 7th February 1997. Although her start was not very lucky, being marred by an engine breakdown and a fire on board, her popularity, her size and her on-board atmosphere, similar to that on the *Fairstar*, should guarantee the vessel a successful new career in Australian waters.

FAIRWIND (1972-1988) - ALBATROS (1993)

(formerly *Sylvania*, then *Sitmar Fairwind, Dawn Princess, Albatros*)
Builders: John Brown & Co. Ltd (Glasgow)
21985 grt (12113 nrt) 7356 dwt; 1993: 24724 grt (13805 nrt) 7356dwt
185.40 [173.74] x 24.40 x 8.94 m
(608.3 [570.0] x 80.1 x 29.3 ft)
2 sets of Pametrada DR steam turbines; 24500 SHP; 20 kn
by builders
1972: 884 passengers, 470 crew
1993: 906 passengers, 500 crew

In the early 'seventies, after the quick decline of passenger transport by sea, Sitmar Line was transformed into Sitmar Cruises and their new *Fairsea* and *Fairwind*, originally intended for the emigrant route to Australia, entered service as full-time cruise ships, being the first vessels in the company history never to operate any liner voyage.

Fairsea and *Fairwind* had almost parallel lives, not only under Vlasov ownership, but also during their previous career.

The eventual *Fairwind*, yard no. 700 of Messrs John Brown Shipbuilding & Engineering Co. Ltd, was launched on 22nd November 1956 by Mrs Robertson, wife of the Canadian High Commissioner in London, who christened her the *Sylvania*. She was the last vessel of the new Cunard Canadian quartette after her sisterships *Saxonia, Ivernia* and *Carinthia* (the eventual *Fairsea*). The four sister-

The *Fairwind* tendering her passengers to St. Croix, Virgin Island on 13th June 1984; the blue stripe at the level of the main deck, introduced for the newbuilding *Fairsky*, had been added the previous month.
(Alberto Bisagno, Genoa)

Fairwind (1972-1988)

*The Sylvania docked at a New York pier on the icy Hudson river in January 1961.
(Victor Young & Len Sawyer, Wellington)*

ships were the first Cunard newbuildings fitted with Denny Brown retractable fin stabilisers.

Intended for the regular weekly service between Liverpool and Montreal, the *Sylvania*'s North American ports were changed to Halifax and New York in winter months when ice made it impossible to steam up the St. Lawrence river.

On 5th June 1957 the *Sylvania*, under the command of Capt. F. G. Watts, sailed on her maiden Atlantic crossing to Montreal. Her first departure was from Greenock instead of her homeport of Liverpool owing to a strike of port workers. In April 1961 the steamer was transferred to the Liverpool-Cobh-New York run, partnering her smaller fleet-mates *Media* and *Parthia* in substitution for the *Britannic*; this last White Star Line motorvessel had in fact been withdrawn from service and sold for scrap.

From 29th May 1964, to increase the number of passengers bound for Europe, the *Sylvania* started to call at Boston, becoming the last Atlantic liner operating from that American port.

Fairwind (1972-1988)

In January 1967 the *Sylvania*'s hull was repainted white for a series of Cunard experimental cruises from Gibraltar.
(Alex Duncan, Gravesend)

The following 23rd December she went back to her builders for a refit of the passenger accommodation, re-entering service on 10th February 1965. Three months later, en route from Liverpool to New York, she altered course to hurry to the aid of the *Lionne*, a Norwegian freighter adrift in a severe storm. 25 of the *Lionne*'s 27-strong crew were rescued while their ship was quickly sinking.

On 24th November 1966 the *Sylvania* sailed for the last Atlantic crossing to New

Waiting to be towed to Trieste for conversion, the *Fairwind* (foreground) and *Fairland* sport the Sitmar colours on their distinctive Clydebank domed funnels.
(Alberto Bisagno, Genoa)

Fairwind (1972-1988)

The Fairwind in her smart new livery, moored at the Trieste Maritime Station, ready for her re-delivery to Vlasov.

York from Liverpool, being transferred upon her return to the Southampton-Montreal line. In January 1967 Cunard chose the ship to inaugurate its experimental season of fly-cruises from Gibraltar and her hull was repainted white. For the occasion an SRN-6 Hovercraft was loaded on board to be used for passengers' excursions at the places visited by the ship.

Different itineraries were offered to the Mediterranean, the Canary Islands, Bermuda and the Caribbean but unfortunately these cruises did not meet the hoped-for response and on 16th May she resumed her liner service to Quebec from Southampton. On 17th June 1967, while steaming off Three Rivers she ran aground and it was not until 7th July that she freed herself and put back to Montreal to be drydocked and checked for seaworthiness. On 20th December 1967 she concluded her last liner voyage in Southampton, sailing two days later on a Christmas and New Year jaunt to Lisbon, Madeira, Tenerife, Casablanca, Gibraltar and Cadiz. She completed another eight fly-cruises from Gibraltar before being laid up in Southampton on 7th May 1968 and quickly put up for sale together with the *Caronia* and the *Carinthia*.

Bought by Sitmar Line on 2nd February 1969 together with the *Carinthia* and registered under the Liberian flag with the new name of *Fairwind*, on 6th January 1970 she left Southampton in tow, arriving twelve days later at the Arsenale Triestino San Marco yard to be re-built into a luxury cruise ship. Original plans by Sitmar were to base their new twin flagships one in Los Angeles and one in Sydney, swapping their roles every six months through a long positioning cruise in the South Pacific.

Like her sistership *Carinthia*, re-named *Fairsea* by Sitmar, the interior of the for-

Fairwind (1972-1988)

mer *Sylvania* was completely stripped down to the steel structure. Also the cargo loading equipment, the front of the superstructure, the funnel and the mast were removed. In the process only part of the original teak covering on the open-air decks, the derricks, the lifeboats (which were disembarked and renewed) and the original power plant were saved. This plant consisted of two sets of high, medium and low pressure turbines by the builders which maintained the 20-knots service speed. Furthermore *Fairsea* and *Fairwind* were the first passenger ships in the World fitted with a biological sewage treatment plant to prevent the pollution of the sea.

Boris Vlasov had in mind a very luxurious and modern cruise ship to enter an ever growing and more demanding market. For this reason Umberto Nordio was called to design the vessel's interiors. He was one of the most famous Italian architects of his time and, among his numerous works, he designed the Trieste Maritime Station and, in the naval field, he designed many interiors of liners fitted out by the Trieste-headquartered Cantieri Riuniti dell'Adriatico, such as the *Homeric*, the *Raffaello*, the *Oceanic* and the post-War refit of the *Conte Biancamano*. *Fairsea* and *Fairwind* were the last projects realised by the great architect who died a few months after they entered service.

At the end of her rebuilding, costing 12.5 million dollars, the *Fairwind* was re-delivered to Sitmar on 14th June 1972, when she sailed under the command of Capt. Rodolfo Potenzoni for a positioning cruise to Los Angeles with many tour operators on board, calling en route at Cadiz, St. Thomas, La Guaira, St. Anna Bay and Acapulco. Actually Sitmar had announced in August 1971 that the *Fairwind* was to be based at Sydney, starting on 18th March 1972 her cruise season of 7-, 14, and 21-day jaunts, with prices per week varying from $170 for an internal standard cabin to $370 for a suite. But while the ship's delivery was delayed by three months there was a re-thinking by the owners and the decision was taken to enter the competitive American market as strongly as possible. This meant that both *Fairsea* and *Fairwind* were to be based in the United States, alternating their cruise

The Promenade Deck Bar of the *Fairwind* and (below) the mural "Caribbean" by San Francisco artist Noal Betts in her Piano Bar; the latter was one of the many pieces of modern art on board the vessel and was made in oils, acrylic and gold leaf.
(William H. Miller, Secaucus)

The Fairwind's Music Lounge.
(William H. Miller, Secaucus)

service on the two American coasts simultaneously. Every effort was made to turn them into competitive vessels. "They are two 'pure' cruise ships and the most luxurious ever seen in most parts of the World", the Vlasov chief executive Giorgio Lauro stated at the time, "The night club goes all night and buffet lunch is served on deck for those who don't wish to go to the dining rooms. There is an average of 50 musicians and entertainers on board at any one time and an Italian crew of 470 to make sure that the best of everything is provided. We print our own colour newspaper on board and the cabins are fitted with telephones from which it is possible to ring any part of the World. Our fares are cheap, at an average of $39 per day, when you consider that normal average fares in the market are $62 a day for the same things. The Sitmar Line has gone to a great deal of trouble to ensure that its new ships are not only abreast of the times, but a few years ahead". The long and successful career of the two vessels, still in operation at the time of writing, proves that they actually were two of the best conversions ever carried out.

Fairwind (1972-1988)

A close up of the aft lido decks and swimming pools.
(William H. Miller, Secaucus)

On 14th August 1972 the *Fairwind* cleared Los Angeles on her first cruise with fare-paying passengers to the Mexican Riviera. On 16th November she weighed anchor from the port of San Francisco for her first trans-Panama Canal cruise to the Caribbean. Like the *Fairsea*, for the following 16 years the *Fairwind* alternated her itineraries to the Mexican Riviera, to the Caribbean and to Alaska, using Los Angeles, Port Everglades and San Francisco as terminal ports, respectively.

On 12th October 1977 *Fairwind* and *Fairsea* met in the Pacific Ocean, some 1500 miles south-west of Los Angeles, during a special "Voyage to Darkness" (as it was billed at the time by Sitmar Cruises) on the occasion of a total eclipse of the sun. The sisterships had on board 1700 passengers, many of them scientists, astronomers, amateur and professional photographers who used the vessels as a privileged observation base for the special astronomic event. It was actually, according to Dr Joseph Chamberlain of Chicago's Adler Planetarium, "a most magnificent spectacle"; luckily the ocean was very smooth and the ships' engines were stopped for a few hours to avoid vibrations disturbing the many telescopes on board.

In June 1987 the *Fairwind* was officially registered with the name *Sitmar Fairwind* and emerged from a refit with a new colour scheme and funnel logo, specially designed for the newbuilding *Sitmar FairMajesty*, which was to be later added to all Sitmar's cruise ships. However the new livery had a short life owing to the 1988 sale of Sitmar's passenger fleet to P&O: in September the *Fairwind* became Princess Cruises' *Dawn Princess*, but her area of operation remained

FAIRWIND (1972-1988)

The *Fairwind* was the only Sitmar vessel to adopt in June 1987 the new Sitmar Cruises corporate identity colour scheme, designed for the newbuilding *Sitmar FairMajesty*, and to have her registered name altered to *Sitmar Fairwind*.
(William H. Miller, Secaucus)

unchanged for the following four and a half years with her new owners. In Summer 1992 the *Dawn Princess* was put on the sale list by P&O, at a price of 32 million US dollars. In a curious twist of fate, she was bought by her former owner, the Vlasov Group, on the following 27th April, becoming the first passenger vessel in its fleet since the Sitmar sale.

In San Francisco on 18th June 1993 the vessel was delivered back to the Vlasov Group which re-named her *Albatros*. The following day she set sail for the South West Marine Inc. yard of Los Angeles, where she was drydocked for a complete overhaul. In the process the Princess logo made way for a new one showing a white albatros flying against a red sun on a blue background.

On 13th July the *Albatros* cleared Los Angeles bound for Bremerhaven, calling en-route at Lisbon for bunkering, crewing and storing. She left Bremerhaven fully booked for her first 10-night cruise to the Norwegian fjords on 18th August, under charter to Phoenix Reisen of Bonn, the German tour operator who have used the *Albatros* until now.

On 27th October 1993 the vessel, after disembarking her passengers at the Genoa maritime station, entered the Mariotti yard for a general revamping of her accommodation which lasted until the following 30th November, when she resumed her cruise service.

Albatros (1993)

The *Fairwind* sailed for Princess Cruises as the *Dawn Princess*.
(Alex Duncan, Gravesend)

On 16th May 1997, the *Albatros* was leaving St. Mary's anchorage in the Scilly Islands when in heavy seas, lead by the local pilot's cutter, she grounded on the rocks. It was the most serious accident during her 30-year life; many breaches were opened in her keel, for a total length of about 60 metres. The divers stated that her underwater body was "peeled back like a sardine tin".

In November 1993, while the vessel is drydocked in Genoa, the already over-painted name *Dawn Princess* is finally unwelded and replaced by *Albatros*.
(Paolo Piccione, Genoa)

Albatros (1993)

Harry's Club on the *Albatros*.
(Jonathan Boonzaier, Singapore)

The Lido Deck and the Columbus Lounge on board *Albatros*.
(Jonathan Boonzaier, Singapore)

However, the vessel remained afloat on the ceiling of her double-bottom and was eventually saved thanks to her one-inch thick plates and her strengthened hull, built to withstand the frequent threat of groundings and the iced surface in the St. Lawrence river. After her passengers were tendered ashore by the local ferry *Scillonian* and the fuel oil from the damaged tanks was pumped into the tanker *Falmouth Endeavour*, the *Albatros* managed to make her way slowly to Southampton where, on 2nd June, she was drydocked for the replacing of 110 tons of hull plates. She left Southampton on 9th July and on the following 19th she was back to service. The *Albatros* offers many different itineraries: in

ALBATROS (1993)

December she usually sails for a round-the-world four-month cruise, while in Spring and Autumn she is mainly employed in Mediterranean cruises; the Summer season sees her based in Bremerhaven for 11-day jaunts to the fjords and Greenland.

Her long-established success and that of her sister ship is a most deserved tribute to their builder and to Boris Vlasov and his staff for having planned their great conversion.

A recent view of the *Albatros* steaming in the Panama Canal; well-maintained and beloved by her crew and by her V.Ships technical managers, she has been a protagonist of many historic stages in the Vlasov Group history.

FAIRSKY (1979-1982)

(never entered service for Sitmar)
(formerly *Príncipe Perfeito*, then *Vera*, *Marianna IX*, *Marianna 9*)
Builders: Swan, Hunter & Wigham Richardson Ltd (Newcastle-upon-Tyne)
19393 grt (6682 nrt) 8630 dwt; 190.50 [172.3] x 23.90 x 7.70 m
(625.0 [565.2] x 78.4 x 25.3 ft)
2 DR set of Parsons steam turbines; 24600 SHP; 20 kn
by builders
1000 passengers in three classes; 200 troops; 320 crew (as *Príncipe Perfeito*)
intended to accommodate abt. 880 passengers and 470 crew for Sitmar Cruises

In 1982 Sitmar's *Vera*, was delivered to her new owner, the Greek John Latsis; it was previously intended to transform her into the cruise ship *Fairsky*, in a similar manner to *Fairsea* and *Fairwind*, but the project never materialised.
(Antonio Scrimali, Alpignano)

In 1975, thanks to the consolidated success of the *Fairsea* and *Fairwind*, Sitmar Cruises decided to add a third ship to work alongside them in the American market. Considering the success of the conversion carried out on the two former Cunarders, they looked for another vessel with similar dimensions to be converted to the same standard.

Perama Bay, the most famous second-hand ship show in the World, offered at the time a vessel suitable for Sitmar's need, the *Queen Anna Maria*.

A technical commission from the Monaco headquarters visited the ship and decided for her purchase. However, when an offer was made to her owner, it was bitterly discovered that the liner had been bought a couple of days before by Carnival to become their *Carnivale*.

Search re-started and in 1978 Sitmar's attention turned to the *Príncipe Perfeito*, which was bought in April 1979 and registered under the Panamanian flag for their Fairline Shipping Corp. with the name *Fairsky*, third vessel in the history of the Group to bear this name.

The *Príncipe Perfeito* was laid down in the Swan, Hunter & Wigham Richardson

Fairsky (1979-1982)

shipyard in Newcastle-upon-Tyne on 12th August 1959 to the order of the Portuguese Companhia Nacional de Navegação and was launched on 22nd September 1960. Built with the contribution of a state subsidy, she was fitted to carry also 200 troops. On 31st May 1961, after reaching 22.5 knots during her sea trials, the *Príncipe Perfeito* was delivered to her owners and on 27th June cleared Lisbon on her maiden voyage to Cape Town, Lourenço Marques, Beira and Mozambique. She was a nice vessel, with a well-balanced profile and one of the most modern and luxurious liners ever to serve the African line.

She remained on her intended route until January 1974 when she was laid up in Lisbon after that the government subsidy for her service was suspended. On the following 25th April she was re-activated for a series of round-voyages to Angola which terminated on 14th June 1975. These proved to be her last commercial voyages: the following month she started a new activity as accommodation ship in Lisbon for repatriated civilians from Africa, an omen of the eventual final employment of the liner. In April 1976 she was sold to Global Transportation Inc. of Panama and, after some refitting carried out in her builders' yard, in June she was permanently berthed at Damman with the new name of *Al Hasa* to give accommodation to 820 local workers. She continued this task until her sale to the Vlasov Group.

In June 1979 an official press release by Sitmar Cruises president John P. Bland disclosed the company's intention for their new acquisition: "The *Fairsky* will join the *Fairsea* and *Fairwind* in early 1981 after a complete rebuilding at esti-

In 1979 A. Storace (better known as Astor), author of advertising material for many shipping companies, produced this picture for the new Fairsky postcard.

The Príncipe Perfeito, eventually Sitmar's Fairsky, leaves Lisbon on 27th June 1961 on her maiden voyage to Cape Town, Lourenço Marques, Beira and Mozambique. (Luís Miguel Correia, Lisbon)

Fairsky (1979-1982)

The Marianna 9 is presently anchored in Elefsis Bay and her future remains uncertain; although she sports the beautiful (and quickly disappearing) lines of a classic liner and seems very well maintained, she is almost 40 years old and her last "finished with engines" was sixteen years ago. (Antonio Scrimali, Alpignano)

mated cost between $40 and $45 million... We are convinced that reconstruction is the best way to go at this time and this acquisition culminates an international search for a suitable existing hull. This $40 to $45 million start-up will enable us to put a first-class luxury vessel in operation in the North American cruise market keeping consumer prices in affordable and competitive ranges".

A call for tenders was made and, among different shipyards interested in the project, the French Chantiers de l'Atlantique, the Italian INMA, the German Lloyd Werft and the Spanish Astilleros Españoles of Barcelona made bids for the *Fairsky*'s reconstruction. At the end the job was awarded to the Spanish shipyard but the latter, on Christmas eve of 1979, asked for the resolution of the contract, after that the feasibility studies showed that the complexity and the cost of the project had been somehow underestimated.

Re-named *Vera*, the former *Príncipe Perfeito* was put up for sale and in June 1982 was bought by Bilinder Marine Corporation, a company controlled by the Greek shipowner John L. Latsis. On the 30th of that month she was moved to Jeddah where, with the name *Marianna IX* (slightly modified to *Marianna 9* in 1984), she resumed her service as accommodation ship for oil crews. In this guise she was later used in the port of Rabegh and in September 1986 she was chartered to the Greek government to accommodate the earthquake victims of Kalamata. The *Marianna 9* was later laid up in Elefsis Bay and, exception made for a further use in Summer 1995 after another earthquake in Egion, she remains idle at anchor in that bay.

FAIRSKY (1984-1988)

(then *Sky Princess*)
Builders: Constructions Navales et Industrielles de La Méditerranée (La Seyne)
46314 grt (22120 nrt) 7673 dwt; 241.00 [203.00] x 27.80 x 7.30 m
 (790.0 [665.8] x 91.1 x 24.0 ft)
2 sets of DR steam turbine; 29500 SHP; 21.8 kn
by General Electric (Lynn)
1600 passengers; 543 crew

After the project of transforming the *Príncipe Perfeito* into the cruise ship *Fairsky* was finally abandoned and no other existing hull satisfying the Sitmar needs was found, Boris Vlasov and his technical staff decided to face the feasibility study for a new vessel to be built ex-novo.

From this decision the fourth *Fairsky* was born. She was the first brand new liner ordered by Sitmar and probably one of the most modern and technologically-advanced vessels of her time, exception made for the decision to engine her with a traditional steam plant when for many years diesels had replaced turbines in passenger ships. The decision stemmed above all from the fact that at the time all the Sitmar fleet was turbine-driven and Sitmar engineers were accustomed and trained to operate and run this kind of propulsion.

In November 1980, a contract for the 46,000 tonner was placed with the French shipbuilder CNIM of La Seyne.

The fourth *Fairsky* in Sitmar's history was also the first passenger newbuilding ever ordered by Vlasov; the photograph depicts her during a call in Vancouver on 21st August 1987. *(Victor Young & Len Sawyer, Wellington)*

Fairsky (1984-1988)

The last and spectacular traditional launching of a large passenger ship took place on 6th November 1982, when the *Fairsky* left the slipway of a French shipyard in La Seyne.
(Rodolfo Potenzoni, Genoa)

The keel of the new flagship was laid on 15th July 1981 on the main slipway of the shipyard and in less than 16 months the hull was completed up to the highest deck, atop of the wheel-house. On 6th November 1982 the *Fairsky* was one of the last passenger ships to be traditionally launched and, with the large complex of the upperworks already in place, the occasion was a rare, spectacular event. While the technical design and the project concept co-ordination were the work of Boris Vlasov himself and his talented newbuilding managers, Arnold Brereton

The *Fairsky*'s fitting out was delayed by long and painful strikes by the local workers, protesting against the announced closure of their yard for lack of work.
(William H. Miller, Secaucus)

FAIRSKY (1984-1988)

On 5th May 1984 in Los Angeles Mrs Jenny Ueberroth addresses her salutatory words to the public as godmother to the brand-new *Fairsky*; sitting close to her are, from left to right, Boris Vlasov, Mauro Terrevazzi, Capt. Rodolfo Potenzoni and his wife; the owner, to honour his origins, wanted an orthodox priest to bless the vessel: he is visible on the extreme left. (Rodolfo Potenzoni, Genoa)

and Ken Norman, four architects were selected and allocated particular accommodation areas of the vessel.

For the suites and standard cabins it was decided that, as the ship would be carrying mainly American passengers, the decorative schemes should be undertaken by Barbara Dorn Associates of San Francisco while responsibility for the public lounges was divided between Dennis Lennon & Partners of London (design director for the *QE2*), Giacomo Erasmo Mortola of Genoa and Clas Olof Lindqvist of Helsinki, with meetings held with all designers present in order to agree upon the transition between the areas of different styles.

The US$ 150 million *Fairsky* ran her trials in early April 1984 and on 2nd May she completed her delivery crossing to Los Angeles under the command of Capt. Rodolfo Potenzoni.

Three days later, during a quite solemn ceremony, Mrs Jenny Ueberroth, wife of the president of the Los Angeles Summer Olympic Committee, officially chris-

Two of the modern public lounges on board the *Fairsky*.

The precious four inches thick Burma teakwood planks being laid on the walkways of the Fairsky's *Promenade Deck, while the first coat of white paint is also being applied.*
(William H. Miller, Secaucus)

tened the vessel, smashing the traditional bottle of champagne on her bow. After the press visit, the new flagship sailed on the very same day for the first of three introductory cruises to the Mexican Riviera, before steaming up the West Coast to San Francisco where she docked for the first time on 6th June 1984.

Re-named *Sky Princess* after the Sitmar buy-out by P&O she continues to alternate Summer cruises from San Francisco to Alaska with Winter jaunts to the Mexican Riviera, using Los Angeles as her terminus port.

The present aspect of the vessel as Sky Princess *in a 6th March 1994 photograph taken in Miami.*
(Andres Hernandez, Miami)

SITMAR FAIRMAJESTY (1989)

Never entered service for Sitmar
(then *Star Princess, Arcadia*)
Builders: Alsthom-Chantiers de l'Atlantique (St. Nazaire)
63524 grt (32185 nrt) 5450 dwt; 246.60 [201.00] x 32.30 x 7.70 m
(809.0 [659.4] x 106.0 x 25.3 ft)
Diesel-electric plant: four 8-cyl. Diesel-generators, 9720kW each;
2 12MW electric motors driving FP propellers; 19.5 kn
by MAN-B&W (Augsburg) - Cegelec (Belfort)
1600 passengers; 543 crew

After their positive experience with the new *Fairsky*, Sitmar immediately embarked on an ambitious newbuilding programme to meet the booming demand in the American market and to combat the strong competition from other companies, all committed to the construction of large and modern cruise ships.

In 1984 negotiations between Sitmar Cruises, Fincantieri and Chantiers de l'Atlantique went on almost simultaneously, but the latter yard, with its great experience gained with successful ships such as the *Nieuw Amsterdam*, *Noordam* and *Sovereign of the Seas*, was able to give more warranties than the Italian company, which had not delivered any significant passenger ship since the mid-'sixties. For this reason the French-built ship, the eventual *FairMajesty*, was the first of the three intended new-buildings to be ordered and completed.

The 63,000 tonner was contracted on 25th June 1986, with an option for a sister ship not exercised, and the first element of the keel was laid twelve months

The *Sitmar FairMajesty* never entered service for Sitmar Cruises, having been completed as *Star Princess*. However, she was floated out on 28th May 1988 and towed to the fitting out berth with her intended name
and the new short-lived Sitmar logo on her funnel.
(Peter C. Kohler, Washington D.C.)

SITMAR FAIRMAJESTY (1989)

One of several preliminary studies specifically realised for the newbuilding *Sitmar FairMajesty* and which was intended to be applied later to all her fleetmates.
(Peter C. Kohler, Washington D.C.)

later. Her name was selected by the Sitmar board of directors from an array of suggestions put forward in a competition by the company's employees. Actually, she was to be officially registered as *Sitmar FairMajesty*; in fact at the time it was decided to add the company's name as a prefix to all the Sitmar ships, in conjunction with the introduction of the new funnel logo and hull colour scheme.

Boris Vlasov, her creator, had died seven months before she was launched by floating out, on 28th May 1988, and the deal to sell Sitmar Cruises to P&O was well advanced; no doubt, *FairMajesty* and her Fincantieri-built derivatives, were one of the salient attractions of the $210 million purchase offer by the British company.

The procedure for the development of the arrangements of the *FairMajesty* was similar to that of the *Fairsky*, but there also were important technical differences, first of all in the selection of the propulsion system. While the *Fairsky* was the last passenger ship fitted with a traditional steam plant, *FairMajesty* was propelled by a diesel-electric system. Although a double set of steam turbines was considered at an early stage of design for both the French and the Italian newbuildings (the eventual *Crown Princess* and *Regal Princess*), it became quickly apparent that it was becoming more and more difficult to locate reliable turbine builders, suppliers of spare parts and repairers and junior engineers with experi-

This model of the *Sitmar FairMajesty* shows not only her final aspect but also the chosen colour scheme although, of course, the blue and red stylised waves were supposed to be painted only over the waterline.
(William H. Miller, Secaucus)

Sitmar FairMajesty (1989)

ence in running and maintaining the steam plant. Boris Vlasov was particularly concerned about noise and vibration problems with internal combustion engines and diesel-electric propulsion provided an obvious compromise, permitting the engines to be fitted on flexible mountings and to run constantly at their ideal speed with the lowest emission of exhaust and vibration.

The positive experience with the conversion of *Queen Elizabeth 2* from steamer to motorship definitely contributed to the adoption of the same solution on *FairMajesty*, which was actually fitted with the same model of engines used for the Cunarder.

Another evident difference between *Fairsky* and *FairMajesty* was the three-deck high lifeboat recesses on the sides, a modern safety solution nowadays compulsory. With regard to the architectural design of the public areas it was decided, in view of certain criticisms of the *Fairsky*'s disparate styles, to use only one company, the Welton Becket Associates Organisation of Los Angeles.

Furthermore, Sitmar Cruises decided to commission alternative studies for a new funnel logo and hull colour scheme which, in the event, would have had an all too brief existence owing to the P&O take-over.

In less than two months from her launching, *FairMajesty* was re-named *Star Princess* and her new Sitmar logo was replaced by the famous Princess Cruises' woman with her hair blowing in the wind.

After her sea trials, during which she reached almost 23 knots on the measured mile (obtaining also an exceptionally low level of noise and vibration), *Star*

The *Star Princess* on her delivery trials off St. Nazaire in February 1989. *(Chantiers de l'Atlantique, St. Nazaire)*

Sitmar FairMajesty (1989)

In December 1997, the *Star Princess* was refitted and transferred to Princess Cruises' parent company with the traditional P&O name of *Arcadia*, as a replacement for the *Canberra*; in this new guise she is here seen during a May 1998 call in Monaco Monte Carlo.

Princess was delivered on 4th March 1989 to P&O. Her proud new owners dedicated a souvenir commemorative book to her, acknowledging the importance of this fine vessel for Princess Cruises. She represented a considerable coup which suddenly strengthened the company's presence in the market, above all enabling them to dramatically shorten the timescale required to introduce new tonnage.

After the *Canberra*'s withdrawal, *Star Princess* was selected as a replacement. She was transferred from the Princess Cruises subsidiary to P&O and, after a revamping of her accommodation in December 1997, she was re-named *Arcadia* and her massive funnel was repainted in the P&O plain buff colour. Her public lounges were indeed decorated with works of art and other memorabilia coming from her famous predecessor.

Presently employed exclusively in the British market, she offers to her passengers 90,000 square metres of accommodation, consisting of a wide range of well furnished public lounges, 14 suites and 36 mini suites with private verandas, and 735 17-sqm standard cabins, 76% of them outside. During her first Summer under the Red Ensign the *Arcadia* was reported to have obtained a high success among her British passengers and this is a considerable achievement considering she took the place of such a beloved vessel as the *Canberra*, rightly considered an institution of British shipping.

Oriana and *Victoria* are the two other partners of *Arcadia* for P&O Summer cruises from Southampton.

CROWN PRINCESS (1990)

(never entered service for Sitmar)
Fincantieri Cantieri Navali Italiani S.p.A. (Monfalcone)
69845 grt (34907nrt) 6995 dwt; 245.00 [204.4] x 32.25 x 7.80 m
(803.8 [670.6] x 105.8 ft)
Diesel-electric plant: 4 Diesel-generators, 9720kW each;
2 12MW electric motors driving FP propellers; 19.5 kn
MAN-B&W (Augsburg) - Cegelec (Belfort)
798 passengers; 656 crew

After the contract with Chantiers de l'Atlantique for the building of the *Sitmar FairMajesty* was signed, the deal with the Italian Fincantieri nevertheless went ahead.

In March 1984 Mauro Terrevazzi and Arnold Brereton, V.Ships' chairman and senior newbuilding project manager, respectively, met senior naval architects Enrico Buschi and Gianfranco Bertaglia from Fincantieri to start a profitable dialogue that would lead, in December 1985, to the signing of a letter of intent for two 600-cabin cruise ships.

Although the eventual *Crown Princess* and *Regal Princess* would be launched after the death of Boris Vlasov and the consequent take-over of Sitmar Cruises by P&O, the design had already been completed by the Vlasov Group and they actu-

Although completed more than two years after his death, the *Crown Princess* could rightly be considered Boris Vlasov's crowning achievement as a marine designer.
(Fincantieri, Trieste)

CROWN PRINCESS (1990)

ally were two Sitmar vessels in every detail, Boris' crowning achievement and a floating tribute to his personal skill as a marine designer and to the brilliant expansion of the Group under his guidance.

It is worth reporting here Mr Brereton's memories of the active role of Boris Vlasov as a passenger ship designer. "Mr Vlasov took a very active part in the design definition, the technical specification and design development of all the vessels ordered by Vlasov Group. Although his formal studies were as an electrical engineer, his experience extended into all aspects of ship design and engineering. From his long association with passenger vessel operation he developed very specific design requirements for this type of vessel to ensure the highest levels of safety and reliability of operation. To this end he was always insistent on using only the best of materials, equipment by the most reputable of manufacturers of more than adequate capacity, and, most importantly, simple and effective system designs well within the capability of the operating personnel.

Mr Vlasov, whilst always erring on the side of caution, was not afraid to lead in the adoption of new materials, equipment or systems which would enhance the safety, quality and reliability of his vessels.

Such decisions were not, however, taken lightly and all aspects of such equipment were carefully analysed and very often improved to his personal satisfaction, before being introduced into the vessels of his fleet".

Fincantieri, whose last passenger ship had been the 1966 *Eugenio C.*, no doubt

The preliminary model and profile (below) presented by Fincantieri to Sitmar in late 1984, was somehow an improved and enlarged version of *Fairsky*.

Renzo Piano revolutionised the ship's skyline and exterior details, although some of his avant-garde proposals (evident in this drawing) were rejected as too costly and complicated to realise; some critics would later define the *Crown Princess* Piano's "halved-ship".

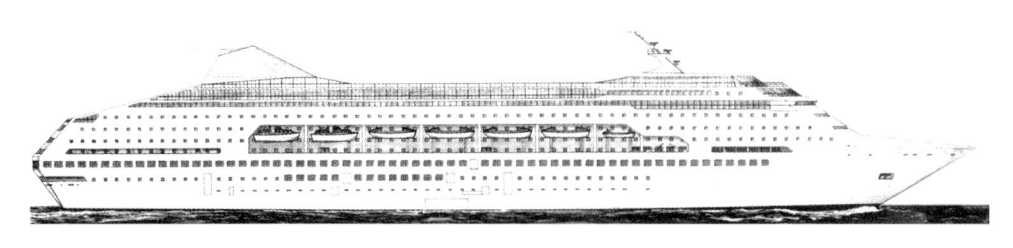

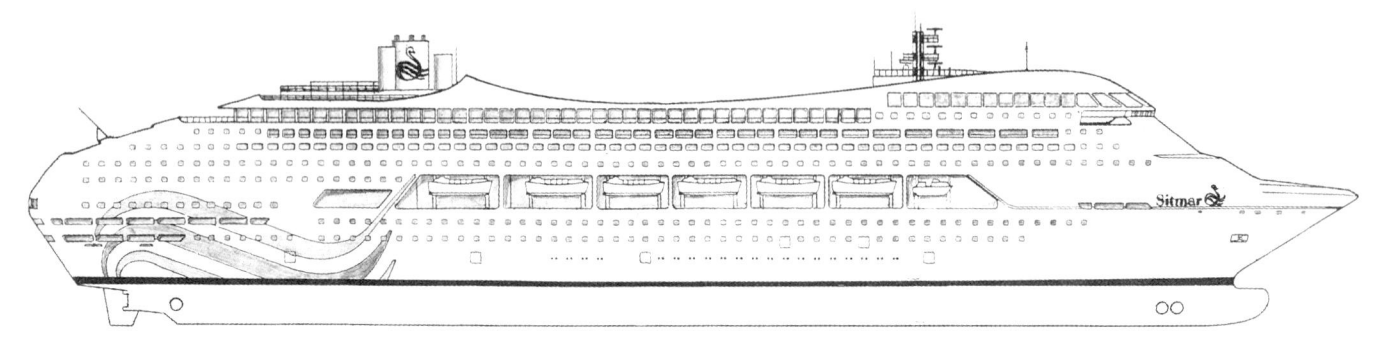

146

CROWN PRINCESS (1990)

Fincantieri's giant bridge-crane lifts into position the pre-fabricated section containing the Crown Princess's casino on 16th May 1989.
(Fincantieri, Trieste)

feels a debt of gratitude to Sitmar and its owner for the confidence he showed them, playing a fundamental role in assuring their present leadership in the building of modern cruise vessels.

On 9th July 1984 Sitmar newbuilding department sent to Fincantieri the preliminary specification of the new ship, with a general arrangement plan realised on the basis of the *Fairsky*'s lay-out which was the reference vessel.

The design of the hotel area of the ship was the work of the Welton Becket Organisation of Los Angeles, the same interior designers used for *Sitmar FairMajesty*. In addition, it was thought that for the new sister ships a fresh approach should be made to colour schemes and material selection and thus another American company, Chambers of Baltimore, were commissioned to work in association with Welton Becket to produce alternative colour and material selections.

Furthermore, as the two vessels were the first major passenger ship contracts to be placed in Italy for some twenty years, it was decided to introduce the renowned Italian architect Renzo Piano to give a touch of Italian style to the newbuildings. On the following 19th December Fincantieri presented to Sitmar the general arrangement plan, technical specification, midship section and machinery arrangement (at the time a traditional steam turbine plant); in particular they had met the owner's request for no less than 85% of outside cabins, against the maximum of 65% on ships of such type in service at the time. These plans were

CROWN PRINCESS (1990)

actually accepted by Vlasov as well as the builder's offer and the design proceeded. In March 1985 it was decided to substitute diesel-electric plant for the steam turbines originally specified and in July a deck was added to the ship, increasing the number of standard two-berth cabins from 600 to 700. In September 1985 the up-to-date plans were approved and Astramar S.r.l., an Italian company purposely founded by Vlasov to deal with the building of the vessels and in order to profit from the Italian subsidies, was opened in Palermo.

In October 1987, only a few weeks before Boris Vlasov's death, Renzo Piano was entrusted with the design of the ship's profile and the interiors of the three upper decks, including the open-air lido and swimming pool area. His basic concept was to realise a mono-volume ship with a clean profile evolved from the silhouette of a dolphin, marking a definitive transition from the classic ocean liner to the modern cruise ship: "...he reproduced the head with the dome over the wheel-house, made the eyes with the bridge-wings, rounded all the corners...".

He had actually started the job one year before when the future Italian Prime Minister Romano Prodi (at the time President of the state-controlled IRI, parent

The distinctive aluminium dome over *Crown Princess* bridge just put in position and the interior of the "dolphin's head" finished. *(Paolo Piccione, Genoa)*

148

Crown Princess (1990)

institution to Fincantieri) personally called Renzo Piano to design a prototype for a modern cruise ship. In five months its plans and volumetric model were ready and were presented in Los Angeles in March 1988.

The eventual *Crown Princess*, yard no. 5839, was laid down in the Monfalcone shipyard on 7th June 1988, one month before Sitmar, Astramar, and the surveyors team of the owner was officially transferred to P&O.

She was floated out on 25th May 1989 and delivered at the maritime station of Trieste one year later, on 29th June.

The *Crown Princess*, at the time the largest cruise vessel ever built at 70,000 grt, was acclaimed at international level for her innovative appearance and for marking the brilliant re-birth of Italian shipbuilding in the passenger ship field, after more than two decades of oblivion.

The *Crown Princess*'s sister ship, the *Regal Princess*, also originally contracted by Sitmar with Fincantieri, was floated out in Monfalcone on 29th March 1990, ran her trials in the Gulf of Trieste on 5th June 1991 and was delivered to Princess Cruises on the following 20th July.

The *Regal Princess*, sister ship to the *Crown Princess* and also originally contracted by Sitmar with Fincantieri, was delivered in Trieste on 20th July 1991. (Alex Duncan, Gravesend)

MINERVA (1996)

Builders: Okean (Nikolajev) and T. Mariotti (Genoa)
12500 grt (3900 nrt) 1500 dwt; 133.00 [115.00] x 20.00 x 5.75 m
(436.4 [377.3] x 65.6 x 18.9 ft)
2 6-cyl. Pielstik Diesel engines; 9400 SHP; 16 kn
2 Wärtsilä/ABB diesel generators; 4000 kW
390 passengers; 156 crew.

The *Minerva*, delivered in April 1996, was the first cruise ship built to the order of Vlasov Group since their 1988 withdrawal from the cruise market. (Mario Bertolotti, Monaco)

In early 1994 Swan Hellenic Cruises, a subsidiary of the P&O Group, decided to look for a replacement of their only vessel, the old *Orpheus*, on charter from Epirotiki for the past 22 years.

Swan Hellenic specialises in cultural and exploration cruises for the British market, during which expert guest speakers give lectures on board, entertaining passengers on their specific field, usually related to the places visited by the vessel.

An agreement was signed between P&O, V.Ships and the Mariotti shipyard of Genoa for the building of a suitable replacement. The eventual *Minerva* would be built by Mariotti and owned by V.Ships, with Swan Hellenic guaranteeing to

Minerva (1996)

The Okean, *as the* Minerva *was unofficially known at the time after the name of the original Russian builder, is here seen being fitted out at the Mariotti shipyard in Genoa.*
(T. Mariotti, Genoa)

keep her on charter for at least a 4-year period. Mariotti of Genoa has no facilities to build a brand new hull of large dimensions and, instead of designing and ordering the construction of a new hull from a sub-contractor (as happened for the *Silver Cloud* and the *Silver Wind*), an already existing one was found at Nikolajev, the Ukrainian city on the mouth of the Ingul river.

The hull destined to become Swan Hellenic's *Minerva* was laid down as yard no. 001 on 30th March 1989 at the Okean shipyard to the order of the U.S.S.R. state fleet and was intended to become a research vessel, fitted with a stern ramp for handling submersibles that could find and track the routes of Allied submarines; however, after the dissolution of the Soviet Bloc, the building was suspended. After its acquisition by V.Ships, the assembly of the ice-classified hull no. 001 was completed to the main deck and it was launched on 10th December 1984. Five days later it was delivered to its new owner and towed to Genoa, where Mariotti would take care of its completion as a passenger ship.

The new plans were drawn up by the V.Ships newbuilding department under the guidance of the senior naval architect Ken Norman and the senior marine engineers Arnold Brereton and Matteo Parodi. The rebuilding project not only involved the superstructure and the interior accommodation, but also the installation of new auxiliary equipment and plant as well as the substitution of the stern. In the process, one of the two stern-thrusters and one of the two bow-thrusters were removed, while the original couple of Dagestan-Machina high-

Minerva (1996)

The Orpheus Lounge on board the *Minerva*.
(Jonathan Boonzaier, Singapore)

efficiency rudders and the power plant, distributed on two shafts driven by a couple of medium-speed diesel engines, were retained.

A couple of Fincantieri fin stabilisers was installed on board, as well as two shaft alternators, two Wärtsilä/ABB diesel generators, two evaporators and one reverse osmosis plant capable of producing 160 tonnes of fresh water a day. In April 1996

The dining room is a clear example of the classic style offered by the *Minerva* to her passengers.
(Jonathan Boonzaier, Singapore)

Minerva (1996)

The large library of the *Minerva* is one of the public rooms preferred by her cultural cruise passengers. *(Jonathan Boonzaier, Singapore)*

the *Minerva* was completed and entered service, sailing from Genoa on her maiden cruise to Mediterranean ports and London, under the command of Capt. Michael Cavaghan. On 7th June 1996 the naming ceremony of the new cruise ship was performed in London's Docklands by HRH the Duchess of Gloucester. Her 398-berth passenger accommodation is distributed on five decks. On the

The aft lido with swimming pool on board the *Minerva*. *(Jonathan Boonzaier, Singapore)*

Minerva (1996)

The Minerva, just a few days before her delivery in April 1996, is almost complete and in a matter of hours the blue painting of the hull will be completed; Swan Hellenic Cruises' logo adorns her funnel.
(Paolo Piccione, Genoa)

lower A and B decks there are the 152 standard cabins, 96 of which are fitted with a large window, while on the upper Bridge and Promenade decks there are the 18 de-luxe cabins and the 24 suites, 12 with private verandas.

On the main deck there are the 250-seat Restaurant, the Wheeler Bar (named after Sir Mortimer Wheeler, a renowned archaeologist and former Swan Hellenic chairman) and the Orpheus main lounge, while on the upper Bridge deck are to be found the Veranda-Café, facing the aft lido and swimming-pool, and two other social rooms connected with the cultural activity on board the *Minerva*: a large library and a conference room. The interiors of the vessel were created in a sober and classic style by the British architects Jan Wilson and Patrick Reardon, taking into account the suggestions made by Swan Hellenic's many repeater passengers.

With British officers and a mixed Ukrainian and Filipino crew, the *Minerva* is employed during the Summer in North European waters, during Spring and Autumn in the Mediterranean and, in Winter, she sails for the Red Sea, India and Far Eastern waters.

SEVEN SEAS NAVIGATOR (1999)

(formerly *Akademik Nicolay Pilyugin, Blue Sea*)
Builders: Admiralty Yards (St. Petersburg) and T. Mariotti (Genoa)
25000 grt (approximate); 170.60 [150.00] x 24.80 x 6.80 m
(559.7 [492.1] x 81.4 x 22.3 ft)
4 SR 8-cyl. Diesel engines; 14600 SHP; 19.5 kn
by Wärtsilä (Swolle)
540 cruise passengers; 326 crew

At the time of writing a luxury cruise vessel, the latest addition to V.Ships' present fleet, is under construction at the Temistocle Mariotti shipyard in Genoa.
The *Seven Seas Navigator* is expected to enter service in Summer 1999 for Golden Ocean 2, a joint-venture company between Vlasov and Radisson-Seven Seas.
In recent years V.Ships and Mariotti have developed a great skill and ingenuity in fitting out de-luxe cruise vessels from existing hulls, the latest result of this being the cruise ship *Minerva*. The latter is, like the *Seven Seas Navigator*, the transformation of a former Soviet Union "spy" ship.
The eventual *Seven Seas Navigator* was laid down on 12th April 1988 as hull no. 0210 in the Admiralty Yards in St. Petersburg and was launched on 24th August 1991 with the name *Akademik Nikolay Pilyugin*. She was the prototype in the intended Akademik class of three sister vessels (which actually never entered service) and constituted the largest and most modern ship conceived by the Soviet navy to follow the movements of Allied submarines. As the original plans show, the U.S.S.R. navy intended to fit them with powerful satellite antennae to create a control net over the movements of the NATO navies.
In November 1993, after the collapse of the Soviet Bloc, the fitting out of the

A drawing by the well-known architectural firm Yran & Stoorbraten of the intended exterior profile of the *Seven Seas Navigator* when completed in Summer 1999.

Seven Seas Navigator (1999)

Akademik Nikolay Pilyugin was suspended. At the time the upperworks, the bridge, the houses, the funnel and the antennas platforms were already in position, as well as the main engines and the auxiliary equipment.

It was in this condition that the vessel entered Genoa under tow on 5th July 1997 with the new name of *Blue Sea* and flying the St. Vincent & Grenadines flag; later her port of registry was changed to Genoa.

As the vessel was in an advanced state of fitting out, with steel interior partitions, insulation, pipes and cabling almost finished, it took nearly a year to gut the hull, with some 4200 tons of materials being scrapped. In the rebuilding process the stern retractable azimuthal thruster was also removed, while the two bow-thrusters were re-engined.

On 12th May 1998, after the superstructure was razed to the main deck, the *Blue Sea* entered drydock to start her rebuilding, commencing with the new engine seats to house two pairs of Wärtsilä 8L/38 Diesel engines, replacing the original Pielsticks. Each pair of diesels is geared to a shaft, (terminating with a variable pitch propeller) which drives also a 2500 kW generator, supplying the electricity for the on-board services together with another three Wärtsilä-Vasa/Leroy-Somer 6R32 diesel-generators of 2000 kW each.

The new power plant will give the vessel a maximum speed of 20.5 knots, with service speed fixed at 19.5 knots.

The extraordinary design for the difficult transformation of what was actually a warship into a five-star cruise vessel is the work of the senior project managers Ken Norman and Arnold Brereton and of the junior project manager Roberto

The Blue Sea, a former U.S.S.R. navy research vessel, arriving in Genoa on 5th July 1997 to be transformed into the five-star cruise ship Seven Seas Navigator thanks to ingenuous re-building plans prepared by V.Ships newbuilding department.

Seven Seas Navigator (1999)

On 26th August 1987 the old upperworks of the former *Akademik Nikolay Pilyugin* are all gone, except for the funnel, which will also go soon to gain access to the engine room.

Fazi, the three talented naval architects and marine engineers of the V.Ships newbuilding department.

Interior décor has been entrusted to the well-known architectural firm Yran & Storbraaten of Oslo, with Mr Søren Storbraaten being responsible for the design co-ordination.

The passenger accommodation is situated on ten of the fourteen decks of the vessel. To enable the passengers to embark easily at any of the ports at which the vessel will call, three different decks are fitted with shell doors and their own vestibule. In addition, on deck 4, there is the tender embarkation area to bring the passenger ashore while the ship is anchored in a roadstead; the *Seven Seas Navigator* has two 150-seat tenders in addition to four lifeboats.

There are in all 250 suites, all exteriors, 215 of which are fitted with private balconies. Each standard suite has an ample area of 29 square metres and a bathroom fitted with separate tub and shower. Among the private staterooms there are four grand and ten superior suites, with an area of 90 sqm and 40 sqm, respectively.

Normal occupancy is double berth per cabin, but additional beds can bring the maximum passenger capacity to 540, while the crew is numbered 326.

The complex of the public lounges is on decks 6, 7 and 12, exception made for the restaurant, placed on deck 5, and for the Panorama Lounge on deck 11, while the suites occupy the whole of decks 8 and 9 and the forward portion of the upper ones.

On board, the passenger flow develops around the central hall, which rises from deck 4 to deck 12 and

A mock-up of a standard suite on the *Seven Seas Navigator*.

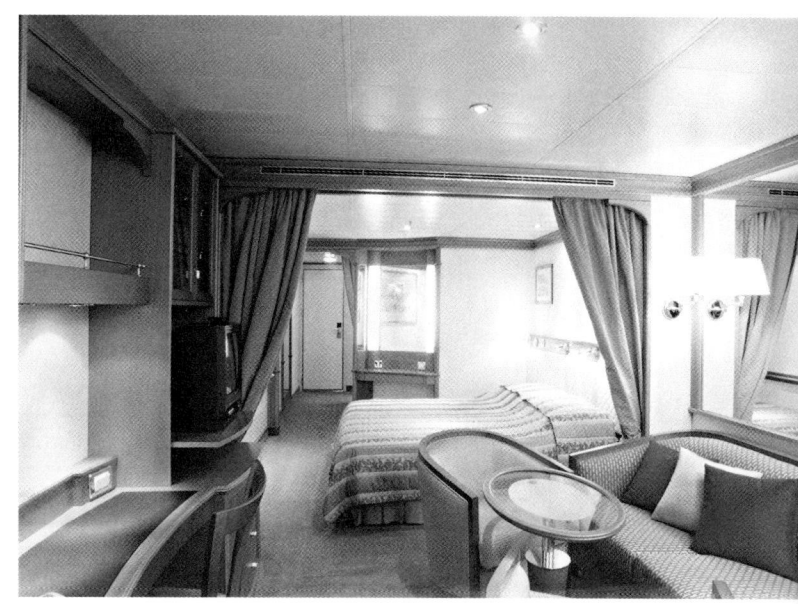

Seven Seas Navigator (1999)

The Blue Sea's hull being gutted at the Mariotti fitting out wharf; note part of the 4200 tons of scrap material on the dock and the historic U.S.S.R. logo still present on her bow.

The Seven Seas Navigator being towed to the drydock for the fitting of the new stern on 16th October 1998.
(Enrico Repetto, Genoa)

is served by three panoramic lifts. Additionally, there is a stern staircase with another two lifts.

Deck 6 contains the main reception area, the card room, the conference room, the cigar lounge "Connoisseur Club", the "Navigator Club" bar, the library and the reading room, while on the upper deck are located the casino and the shops' gallery. From the aft part of the latter deck, passengers gain access to the two-level "Seven Seas" show lounge. A part of the lower deck of the show lounge acts also as night club and disco with its own bar for late bird passengers.

The stern portion of deck 11 is occupied by the "Galileo" Panorama Lounge, a quiet place with piped-in soft music, deck to deck glass walls facing the sea and facilities for early bird passengers' breakfast.

Another pole of attraction for passengers is deck 12, completely devoted to the health centre, and fitted with gymnasium, massage rooms, thalasso theraphy, turkish bath and sauna. At the forward end the fitness centre opens onto the "Vista" Observation Lounge, a relaxing place with a panoramic view similar to that of the bridge, located one deck below. The Vista Lounge is actually fitted with some wheel-house instrument repeaters, such as GPS, map, speed and route indicator and other screens giving meteorological and navigating information.

The "Compass Rose" Restaurant, located on deck 5, offers an open one-seating service with tables ranging from two to eight places. Alternatively, the passen-

Seven Seas Navigator (1999)

gers can choose the 24-hour cabin service (where it is possible to have not only breakfast but also lunch and dinner), the barbecue buffet of the swimming pool or the self-service of the "Portofino Grill" lounge, located on the stern area of Lido Deck.

This room, served by a second independent fully equipped galley, offers 200 in-door seats plus 40 on the stern balcony; during the evening the self-service is closed and it becomes an exclusive "à la carte" restaurant with a beautiful night seascape.

The *Seven Seas Navigator* is one of the very first cruise ships to be designed with the so called "full green concept".

Thanks to her most up-to-date sewage and garbage treatment system, she complies with the most stringent rules for pollution prevention and can sail all over the World, including special protected areas.

She is a very versatile vessel: not conceived for repetitive cruises but to navigate the seven seas, the new cruise ship has a provision capacity of at least six weeks and an autonomy of 7500 miles at cruising speed.

On 23rd December 1999 a contract for the building of a similar cruise ship was signed with Chantiers de l'Atlantique of St. Nazzire; the French shipbuilder will start the costruction in July 1999 and deliver the vessel in Summer 2001.

For the Vlasov Group this newbuilding marks a new, successful achievement in their history of continuous progress: Alexandre and Boris Vlasov would be proud of her and, in a curious twist of fate, she was born like them in Russia to become well known all over the World.

One of the new diesel-generators being lifted on board.
(Roberto Fazi, Milan)

Tankers, ore/oil and product carriers

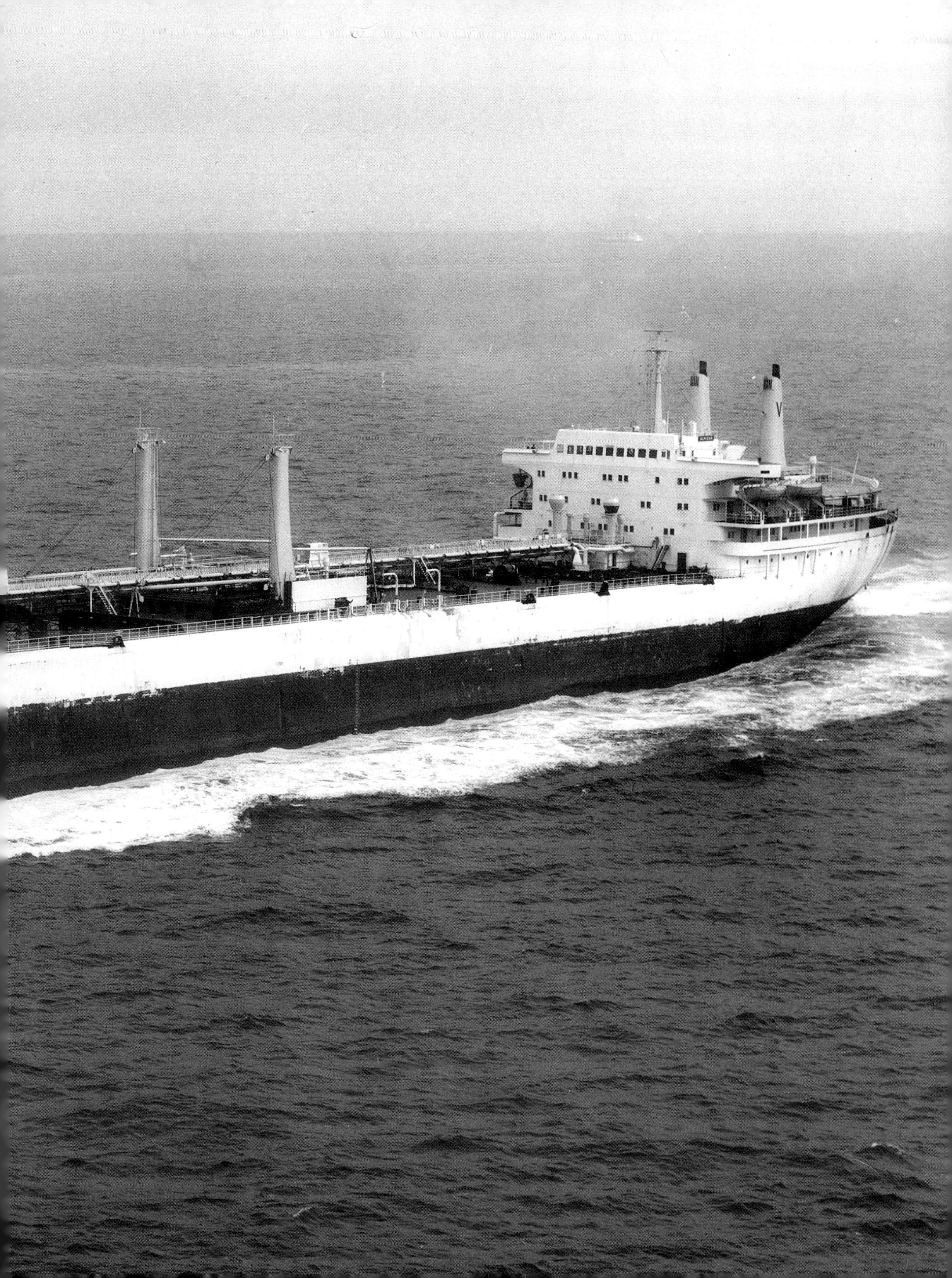

The shipping activities of Alexandre Vlasov stemmed from his business in the coal trades. When, after the Second World War, oil started to replace the use of coal in many industrial processes it seemed a natural move for the Vlasov Group to enter the market, with a brand new fleet of tankers.

In the early 'fifties they placed orders for seven new vessels at various yards, which specialised in the building of oil tankers. The first tanker in the history of the Vlasov Group was the *Almak*, delivered in January 1952 by the famous Clydebank shipbuilder John Brown & Co. She was followed by a sistership the following May and within less than two years another five larger fleet-mates were in service.

The new tankers were, at the beginning, registered under the ownership of the Alvion Steam Ship Corp. of Panama and the Alva Steam Ship Co. Ltd of London, with the Navigation & Coal Trade Co. Ltd acting as managers.

During the first two months of 1966 the tanker fleet, which had grown to twelve vessels, was transferred to subsidiary companies (one for each ship) and placed on the Liberian register.

In 1974, it was decided to replace the older tonnage in the fleet. A newbuilding programme was commenced and five sisterships of around 55,000 dwt each were ordered from the British yards of Cammell Laird. This well-known shipbuilder had developed a new design for a type of standard tanker to cope with the new demands of the market. The prototype vessel, *Algol*, was delivered to Vlasov on 2nd May 1977. This class of vessel, known as StaT55, has proven to be very successful and they all remain in service at the time of writing.

Two important developments took place in the mid-'seventies when the Group concluded its acquisition of a large British shipping concern and, in addition, founded the first Western-owned shipping company to be based in Saudi Arabia. In 1973, Navcot Shipping Holdings Ltd, a joint venture between Vlasov and Capitalfin (a financial company controlled by the Italian Banca Nazionale del Lavoro, Fiat and Montedison) bought the British-owned Shipping Industrial Holdings Ltd, parent company of Dene Shipping Co. Ltd (founded in 1937) and Silver Line Ltd (founded in 1925). They brought to Vlasov a fleet of 25 modern vessels (tankers, product carriers, chemical carriers and bulk carriers) with two newbuildings under construction.

In November 1975 the Amar Line, a subsidiary of the Arabian Maritime Company, was founded under the joint ownership of Vlasov and Saudi Arabian

Tankers, ore/oil and product carriers

entrepreneur Gaith Pharaon. Thanks to its new partner, the Vlasov Group would survive the petroleum crisis of those years. The frightening increase in the cost of the fuel forced many shipowners into serious financial difficulties if not to bankruptcy.

In 1977 Gaith Pharaon sold his 50% holding in the Amar Line; 20% were taken over by Vlasov while the remaining 30% were purchased by Al Rajhi-Al Jedais Company-Riyadh. Amar Line was dissolved in 1985.

In 1977 Vlasov bought out the shares of his partners in Navcot Shipping Holdings Ltd and Shipping Industrial Holdings. Ltd was merged with Navigation & Coal Trade Co. Ltd, with the name Silver Line becoming their trade-mark. Silver Line took on the technical management of the Vlasov-owned fleet, lightening the tasks of the Monaco headquarters which were thus able to concentrate on offering ship management to other shipowners, nowadays the main activity of the Group under the trademark V.Ships.

The general arrangement plan for the 230,000 dwt sister tankers *Alva Star*, *Alva Bay* and *Alva Sea*, the largest vessels ever owned by the Vlasov Group.

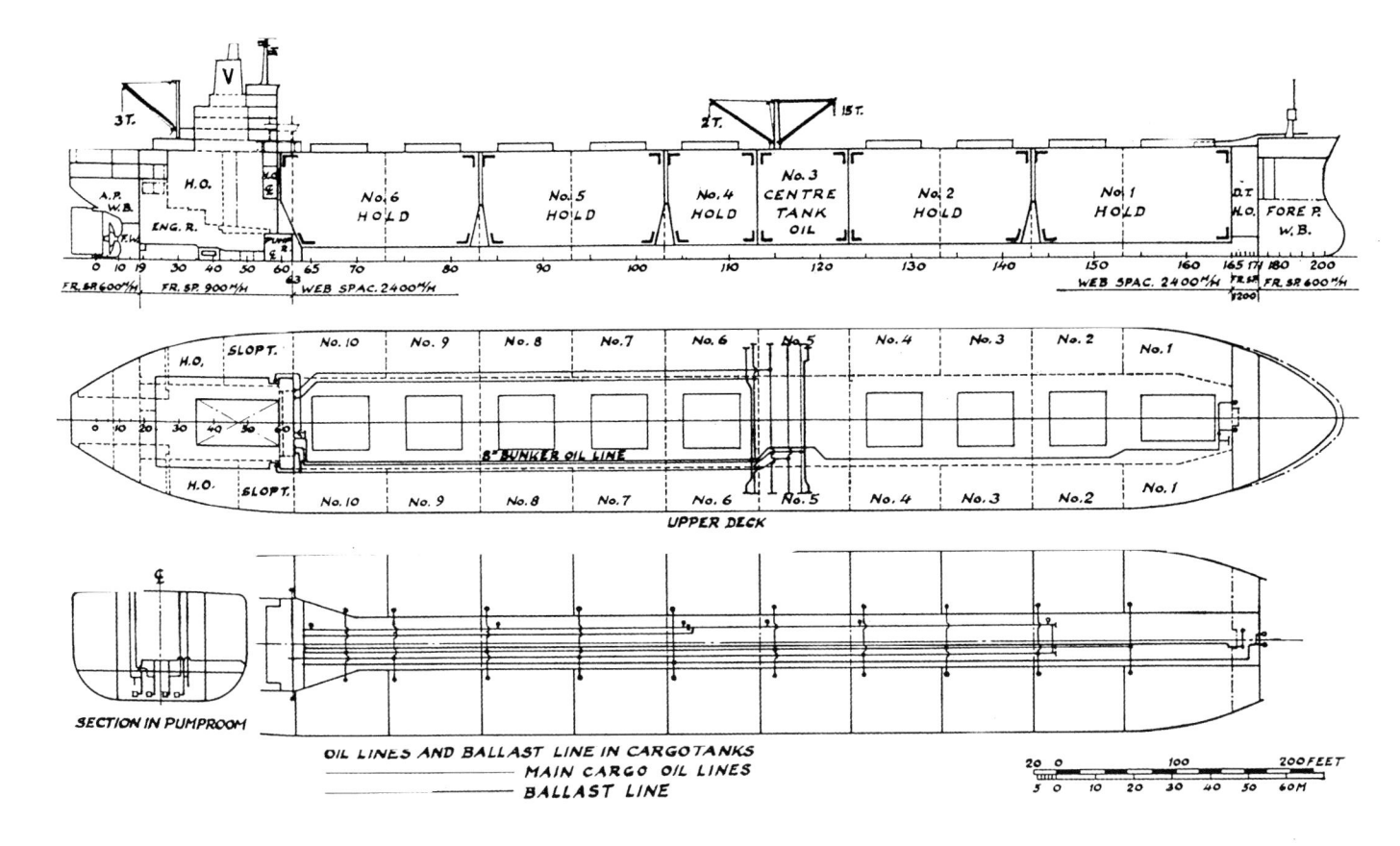

FLEET LIST DATA READING KEY

NAME (YEARS OF SERVICE FOR VLASOV GROUP)
Type of vessel (former names, subsequent names)
Builders: name (place of building)
gross tonnage (net tonnage) deadweight;
length overall [length between perpendiculars] x beam moulded x full load draught
number of engines and type; Brake (oil engines), Indicated (steam reciprocating), Shaft (turbines) or
Nominal HP; service speed
name of engine builders (place where built)

Note: as in the history of the Vlasov Group there are many vessels sharing the same name, a chronological number is given after the ship's name to facilitate her identification. This fleet list contains only the vessels owned by the Vlasov Group and which operated for it. An average of five/six vessels per year is presently bought and sold as part of the company's shipbroking activity; these ships, which do not operate under the Vlasov banner, as well as chartered vessels, are not included.

ALMAK I (1952 - 1971)

Tanker (then *World Hope, Dragon*)
Builders: John Brown & Co. Ltd (Clydebank)
13070 grt (11255 nrt) 7431 dwt
170.38 [162.15] x 22.04 x 12.25 m
(559.0 [532.0] x 72.3 x 40.2 ft)
1 single-acting 6-cyl. Doxford Diesel engine
6200 SHP; 15 kn
by builders

1951 October 18th: launched.
1952 January: completed and delivered to Alvion S.S. Corp; p.o.r. Monrovia.
1966: transferred to Almak Shipping Corp., Monrovia.
1971: sold to Black Navigation Co. Inc. Panama and re-named *World Hope*.
1972: sold to Imahatsu Kaiun K.K., Panama and re-named *Dragon*.
1974 March 7th: arrived at Kaohsiung to be broken up.

ALGOL I (1952- 1969)

Tanker
Builders: John Brown & Co. Ltd (Clydebank)
13070 grt (11255 nrt) 7431 dwt
170.38 [162.15] x 22.04 x 12.25 m
(559.0 [532.0] x 72.3 x 40.2 ft)
1 sungle-acting 6-cyl. Doxford Diesel engine;
6200 SHP; 15 kn
by builders

1952 March 13th: launched. May: completed and delivered to Alvion S.S. Corp.; p.o.r. Monrovia.
1966: transferred to Algol Shipping Corp., Monrovia.
1969 February 10th: caught in a storm while en route from Puerto Cabello to New Bedford; aground in Buzzards Bay; cargo transferred to another tanker, refloated and towed to New York; damaged beyond economic repair; laid up. June 9th: arrived in Hamburg in tow from New York; later sold to Spanish shipbreakers. September 4th: arrived in tow at Santander from Hamburg and delivered to local shipbreakers on the following 17th.

ALVA STAR I (1953-1967)

Tanker (then *Angel Gabriel*)
Builders: Sir James Laing & Sons (Sunderland)
12223 grt (7332 nrt) 18340 dwt
165.64 x 22.23 x 9.35 m (543.4 x 72.9 x 30.7 ft)
1 single-acting 6-cyl. Diesel engine
6200 SHP; 15 kn
by William Doxford & Sons Ltd (Sunderland)

1953 February 16th: launched with the name Alva Star; delivered to Alva S.S. Co. Ltd, London and entered service during the following June.
1958: transferred to Alvada Shipping Corp. Ltd, Hamilton.
1967: sold to Cherouvin Cia. Maritima S.A., Piraeus, Greece and re-named *Angel Gabriel*.
1969 September 23rd: when approaching Malta from Venice in ballast an explosion in the engine room left the vessel adrift; wrecked on the rocks of Siberia Point, broken in two, total loss.

ALKAID (1953 - 1973)

Tanker
Builders: Nederlandsche Dok & Scheepbouw (Amsterdam)
15805 grt (9776 nrt) 23790 dwt
183.82 [173.00] x 23.20 x 9.75 m
(603.1 [567.6] x 76.1 x 32.0 ft)
1 7-cyl. Stork Diesel engine; 7400 BHP; 14 kn
by builders

1953 July: completed and delivered to Alkaid Shipping Corp.; p.o.r. Monrovia.
1973 November 18th: while in tow from Jacksonville to Taiwan to be broken up, she broke the tow line in a storm. November 24th: found aground on Diogo Islands, at 20°42'42"N, 121°56'24"E; she subsequently broke in two and sunk.

Tankers, ore/oil and product carriers

The *Alkaid*, built in 1953.

ALVA CAPE (1953-1966)
Tanker
Builders: Greenock Drydock Co. Ltd (Greenock)
11252 grt (6421 nrt) 16635 dwt
166.74 x 21.17 x 8.90 m (547.0 x 69.5 x 29.2 ft)
1 single-acting 6-cyl. Diesel engine
6200 SHP; 15 kn
by Scotts' S.B. & Eng. Co. Ltd (Greenock)

1953 May 15th: launched. September: delivered and entered service for Alva S.S. Co. Ltd, London.
1966 June 16th: collided with *Texas Massachusetts* in New York Harbour and caught fire. Towed out of the harbour and sunk by gunfire by a U.S. Coast Guard Cutter.

ALVA BAY I (1953-1969)
Tanker
Builders: W. Hamilton & Co. Ltd (Port Glasgow)
11340 grt (6582 nrt) 16509 dwt
166.53 x 21.17 x 8.90 m (546.4 x 69.5 x 29.2 ft)
1 single-acting 6-cyl. Diesel engine
6250 SHP; 15,5 kn
by D. Rowan & Co. Ltd (Glasgow)

1953 August 11th: launched.
1953 October: delivered and entered service for Alva S.S. Co. Ltd, London.
1969 April 24th: arrived at Hirao, Japan to be broken up.

The 1953-built *Alva Star* was the first tanker owned by Alva S. S. Co. Ltd.

Tankers, ore/oil and product carriers

The first *Alvega*, built in 1956.

ALKOR (1953 - 1972)

Tanker
Builders: Nederlandsche Dok & Scheepbouw (Amsterdam)
15805 grt (9776 nrt) 23820 dwt
183.82 [173.00] x 23.20 x 9.75 m
(603.1 [567.6] x 76.1 x 32.0 ft)
1 7-cyl. Stork Diesel engine; 7400 BHP; 14 kn
by builders

1953 November: completed and delivered to Alkor Shipping Corp.; p.o.r. Monrovia.
1972 June 6th: arrived at Bilbao in tow from Bremerhaven to be broken up by Eduardo Varella.

ALVEGA I (1956-1977)

Tanker
Builders: John Brown &Co. Ltd (Clydebank)
21164 grt (13032 nrt) 35427 dwt
202.05 [194.57] x 26.37 x 14.10 m
(662.9 [640.0] x 86.5 x 46.3 ft)
2 double-reduction steam turbines
11500 SHP; 14.5 kn
by builders

1955 September 21st: launched.
1956: completed and delivered to Alvion S.S. Corp., Panama, Panamanian flag.
1966 February: transferred to Alvega Shipping Corp., Monrovia, Liberian flag.
1977 February 25th: arrived at Pusan (Korea) to be broken up by Dong Kuk Steel Mill Co.

ALVENUS I (1957 - 1970)

Tanker (then *Trader*)
Builders: Vickers-Armstrongs S.B. Ltd (Walker-on-Tyne)
24468 grt (13214 nrt) 32682 dwt
202.02 [195.38] x 26.49 x 10.55 m
(662.8 [641.0] x 86.9 x 34.6 ft)
2 double reduction steam turbines
26630 SHP; 15 kn
by builders (Barrow)

1956 August 2nd: launched.
1957 January: delivered to Alvion Shipping Corp. Monrovia and entered service.
1967: transferred to Alvenus Steamship Corp. Monrovia.
1970: sold to Don Shipping Co. Ltd, Piraeus and renamed *Trader*; p.o.r. Piraeus.
1972 June 11th: declared total loss.

ALNAIR I (1964 -1977)

Tanker
Builders: Deutsche Werft AG (Hamburg)
31330 grt (19737 nrt) 60221 dwt
236.20 [228.00] x 32.21 x 16.01 m
(774.9 [748.0] x 105.7 x 52.5 ft)
2 double reduction steam turbines
19800 SHP; 15.5 kn
by Westinghouse Electric Corp. (Philadelphia)

1963 September 9th: launched with the name *Alnair*.
1964: delivered and entered service for Alvion S.S. Corp., Panamanian flag.
1966 January: transferred to Alnair Shipping Corp., Monrovia; Liberian flag.
1977 November 25th arrived at Kaoshiung to be broken up.

ALTANIN I (1964 - 1976)

Tanker
Builders: Deutsche Werft AG (Hamburg)
44935 grt (31554 nrt) 89933 dwt
265.18 [254.01] x 36.61 x 18.29 m
(870 [833.4] x 120.1 x 60.0 ft)
2 double-reduction steam turbines
19800 SHP; 15.5 kn
by Westinghouse Electric Corp. (Lester)

1961: a group of various independent owners (Niarchos, Onassis, Sigval Bergesen, Fred. Olsen & Co., Anders Jahre, Vlasov) ordered nine 90,000 dwt tankers against a 20-year bare-boat charter to

TANKERS, ORE/OIL AND PRODUCT CARRIERS

Shell; the Vlasov tanker was the *Altanin*.
1964 April 11th: launched for Alvion S.S. Corp., Panamanian flag.
1966 January: transferred to Altanin Shipping Corp., Monrovia; Liberian flag.
1976: transferred to Alkor Shipping Co. Ltd, London; Navigation & Coal Trade Co. Ltd managers; British flag.
1978: broken up in Taiwan.

ALMIZAR (1964-1982)

Tanker
Builders: Deutsche Werft AG (Hamburg)
44900 grt (31500 nrt) 89900 dwt
267.93 [254.01] x 36.61 x 18.29 m
(879.0 [833.4] x 120.1 x 60.0 ft)
2 double-reduction steam turbines
19800 SHP; 15.5 kn
by Westinghouse Electric Corp. (Philadelphia)

1964 August 20th: launched for Alvion S.S. Corp., Panamanian flag.
1966: transferred to Almizar Shipping Corp., Monrovia; Liberian flag.
1971: jumboized at Yokohama by Mitsubishi Heavy Industries Ltd by the insertion of new side tanks and forward sections.
New dimensions: 268.92 [255.00] x 43.21 x 19.18 m; 51602 grt (39173 nrt) 110363 dwt.
1976: time-chartered to Shell, trading Arabian Gulf to Far East.
1980: transferred to Andwei Holdings Inc., Monrovia.
1981 December 14th: caught in a storm.
She lost the left side tank, suffering severe structural damage, while en route from Puerto La Cruz to Europort.
December 24th: arrived Setubal.
Inspection showed the vessel to be beyond economic repair; laid up.
1982 June 8th: arrived at Barcelona from Setubal to be broken up by Desguages Cataluna.

ALRAI I - CASPIAN SEA (1967- 1978)

Tanker (formerly *Northern Gulf*)
Builders: Nederlandsche Dok & Scheepbouw (Amsterdam)
24424 grt (15215 nrt) 37750 dwt
212.14 [201.17] x 27.58 x 11.88 m
(696.0 [660.0] x 90.5 x 39.0 ft)
2 double-reduction steam turbines
13750 SHP; 15 kn
by Parsons Marine Turbine Co. Ltd (Wallsend-upon-Tyne)

The *Altanin*, launched in 1964, was part of a group of nine sisterships ordered by different private shipowners for a long-time charter to Shell.
(Antonio Scrimali, Alpignano)

The first *Alnair*, delivered to Vlasov in 1964.
(Alberto Bisagno, Genoa)

TANKERS, ORE/OIL AND PRODUCT CARRIERS

1956 July: completed and delivered to Afran Transport Co., Monrovia, as *Northern Gulf*; Liberian flag.
1967: sold, together with her sisterships *Western Gulf* (re-named *Alkes*) and *Eastern Gulf* (re-named *Alioth*), to the Vlasov Group; re-named *Alrai* and registered to the Alrai Shipping Corp. Monrovia; Liberian flag.
1974: transferred to Black Sea Shipping Co. Ltd, London and re-named *Caspian Sea*.
1978: sold to Taiwanese shipbreakers.

ALIOTH - ALSHARKIAH (1967-1978)

Tanker (formerly *Eastern Gulf*)
Builders: Chantiers et Ateliers de Saint Nazaire (Penhoët)
25168 grt (15534 nrt) 38329 dwt
211.94 [201.17] x 27.95 x 11.78 m
(695.3 [660.0] x 91.7 x 38.6 ft)
2 double-reduction Parsons steam turbines
38220 SHP; 16.5 kn
by builders

1956 May: completed and delivered to Afran Transport Co., Monrovia, as *Eastern Gulf*.

1967: sold together with her sisterships *Western Gulf* (re-named *Alkes*) and *Northern Gulf* (re-named *Alrai*) to the Vlasov Group; re-named *Alioth* and registered to the Alioth Shipping Corp. Monrovia; Liberian flag.
1976: transferred to Amar Line, Jeddah; re-named *Alsharkiah*; Saudi Arabian flag.
1978 June 10th: arrived in Kaoshiung to be broken up by Kwang Steel Enterprises.

ALKES - BARILOCHE (1967 - 1977)

Tanker (formerly *Western Gulf*)
Builders: Nederlandsche Dok & Scheepbouw (Amsterdam)
24424 grt (15215 nrt) 37750 dwt
212.14 [201.17] x 27.58 x 11.88 m
(696.0 [660.0] x 90.5 x 39.0 ft)
2 double-reduction geared turbines
15000 SHP; 15.5 kn
by builders

1956 March: completed and delivered as *Western Gulf* to Afran Transport Co., Monrovia; Liberian flag.

The *Alrai*, formerly *Northern Gulf*, was one of three second-hand sisterships bought by Vlasov in 1967.

Tankers, ore/oil and product carriers

1967: sold together with her sistership *Eastern Gulf* (re-named *Alioth*) and *Northern Gulf* (re-named *Alrai*) to the Vlasov Group; re-named *Alkes* and registered to Alkes Shipping Corp. Monrovia; Liberian flag.

1977 March: re-named *Bariloche*; after a few months of trading in Caribbean and Argentine waters she was laid up. September: delivered to Li Chong Steel & Iron Works, Taiwan for demolition.

ALVA STAR II - AL SAUDIA (1970-1982)

Super-tanker
Builders: A/B Gotaverken (Gothenburg)
113933 grt (90603 nrt) 231759 dwt
332.30 [320.04] x 45.60 x 26.67 m
(1090.2 [1050.0] x 149.6 x 87.5 ft)
2 double-reduction steam turbines
32450 SHP; 15 kn
by Stal Laval Turbines Co. (Finspang)

1970: delivered to Alva S.S. Co. Ltd, London.
1971: transferred to Alva Star Shipping Co. Ltd, London, subsidiary of Alva, British flag.
1975: on time-charter to BP.
1980: transferred to the Amar Line; re-named *Al Saudia*; Shipping Management S.A.M. Monte Carlo managers.
1982 November 25th: arrived at Inchom, South Korea to be broken up by Inchom Iron & Steel Co.

ALCHIBA - ALMA THAR (1972-1986)

Tanker
(then *Alchiba*, *Lyford*, *Cadore*, *Laurotank Cadore*, *Cadore*, *Londontank*, *Xanadu*)
Builders: Stabilimenti Navali S.p.A. (Taranto) and Officine T. Mariotti (Genoa)
17299 grt (11327 nrt) 28769 dwt
192.41 [182.53] x 24.50 x 13.06 m
(631.3 [598.9] x 80.4 x 42.8 ft)
1 single-acting 6-cyl. Diesel engine
12000 BHP; 15 kn
by FIAT Grandi Motori (Turin)

The *Alkes* of 1967.
(Alberto Bisagno, Genoa)

The 1970-built *Alva Star*; all the British registered ships owned by the London-headquartered Vlasov's subsidiary used Alva as prefix to their names.

TANKERS, ORE/OIL AND PRODUCT CARRIERS

In 1986 the former *Alva Bay* (below), at the time called the *Clearwater Bay*, was deliberately sunk at Inchom, South Korea, to facilitate the building of a new port breakwater.

1970: launched as *Alchiba* by Stabilimenti Navali di Taranto S.p.A.; work suspended after the yard's bankruptcy.
1971: towed to Officine Mariotti, Genoa; engined with a Fiat 786S Diesel instead of the intended steam plant; fitting-out completed; in March delivered to Alchiba Shipping Corp., Monrovia and entered service; Liberian flag.
1976: time-chartered to Exxon.
1978: transferred to Amar Line, Arabian Maritime Transport Co. Ltd, Jeddah, Saudi Arabia; re-named *Alma Thar*, Saudi Arabian flag.
1980: transferred to Algol S.S. Co. S.A., Panamanian flag. Re-named *Alchiba*.
1983: re-named *Lyford*, same owner, Bahamian flag.
1986: sold to Sovereign Shipping Enterprises Ltd, Nassau. October: sold to Adriatica Tankers S.p.A. and re-named *Cadore*, p.o.r. Venice.
1988: sold to Starlauro S.p.A. of Naples and re-named *Laurotank Cadore*.
1989: name briefly reverted to *Cadore* and later re-named *Londontank*.
1990: sold to United Resources Group, Malta and re-named *Xanadu*; World Carrier Ltd London managers; p.o.r. La Valletta.
1998: in service.

ALVA BAY II (1973 - 1983)

Ore/Oil super-carrier (then *Clearwater Bay*)
Builders: A/B Gotaverken (Gothenburg)
120698 grt (99783 nrt) 225898 dwt
332.25 [320.04] x 45.60 x 27.34 m
(1090.2 [1050.0] x 149.6 x 89.7 ft)
2 double-reduction Stal-Laval steam turbines
31955 SHP; 15 kn
by builders

1973: delivered to Alva subsidiary Alva Bay Shipping Co. Ltd, London, British flag.
1981: Shipping Management S.A.M. Monte Carlo managers.
1983: sold to Pu Tai Shipping Co., Hong Kong; re-named *Clearwater Bay*.

Tankers, ore/oil and product carriers

1986 June 4th: sold to Inchom Iron & Steel Co. for break up and transferred to the scrapyard at Inchom, South Korea. Before demolition commenced she was re-sold to the local port authority and sunk to act as port breakwater.

ALVA SEA - RED SEA (1973- 1987)

Ore/Oil super-carrier (then *Coral Rose, Bos Combo, Combo, Compass Rose*)
Builders: A/B Gotaverken (Gothenburg)
120708 grt (99956 nrt) 225102 dwt
332.25 [320.04] x 45.60 x 27.34 m
(1090.2 [1050.0] x 149.6 x 89.7 ft)
2 double-reduction Stal-Laval steam turbines
31955 SHP; 15 kn
by builders

1973: delivered to Alva subsidiary Alva Sea Shipping Co. Ltd, London, British flag.
1979 April: re-engined at Yokohama, Japan by Mitsubishi Heavy Industries Co. Ltd; 1 single-acting Diesel engine; 32000 SHP; 16 kn; re-entered service the following August.
1985: transferred to Alva Sea Shipping Co., Gibraltar and re-named *Red Sea*.
1987: sold to the Avalon Maritime Ltd of Gibraltar, re-named *Coral Rose*; Drake Maritime S.A., Gibraltar managers.
1990 May 10th: class withdrawn by Lloyd's Register at owner's request; laid up and re-named *Bos Combo*.
1991: re-named *Combo* and put up for sale; bought by Compass Shipping S.A. and re-named *Compass Rose*; p.o.r. Majuro, Marshall Islands.
1992 August 30th: weather damage while en route from Ponta da Madeira to Kwangyng. September 7th: arrived in Singapore and entered shipyard to be repaired, but work later suspended. November 8th: sold to Chinese shipbreakers.

BERMUDA SEA (1974 - 1977)

Tanker (formerly *Yoho Maru*, then *Grand Youth*)
Builders: Hitachi Zosen (Innoshima)
48810 grt (22102 nrt) 87063 dwt
244.00 [234.00] x 37.11 x 14.61 m
(800.5 [767.7] x 121.8 x 47.9 ft)
1 single-acting 9-cyl. Diesel engine
20700 BHP; 15 kn
by builders

1965: completed and delivered to Lino Kaiun KK, Tokyo, as *Yoho Maru*.
1974: sold to Vlasov's White Sea Shipping Co., Hamilton, a subsidiary of Black Sea Shipping Co. Ltd, London and re-named *Bermuda Sea*; Bermudan flag.
1977: sold to Erskine Shipping Corp. Monrovia and re-named *Grand Youth*.
1983 March 21st: arrived at Pusan, Korea to be broken up by Da Dong Yin Hong Un Co.

ALNAJDI (1975 - 1982)

Tanker (formerly *Halcyon Breeze, Derwentdale*, then *Al Najdi*)
Builders: Hitachi Zosen (Innoshima)
42343 grt (28288 nrt) 74552 dwt
243.52 [232.01] x 35.80 x 13.00 m
(798.9 [761.2] x 117.5 x 42.6 ft)
1 single-acting 9-cyl. B&W Diesel engine
20700 BHP; 15 kn
by builders

1964: delivered to Caribbean Tankers Ltd, London as *Halcyon Breeze*.
1967: bare-boat chartered to Court Line Ltd and re-named *Derwentdale*.
1975: bought by Amar Line, Jedda and re-named *Alnajdi*.
1979: name respelled *Al Najdi*.
1982 May 11th: arrived at Kaoshiung to be broken up by G.I. Yuen Steel Enterprise Co.

ALGOL II - OCELOT (1977 - 1985)

Tanker (then *Dagli, Iver Lundina, Lundina, Santa Lucia*)
Builders: Cammell Laird S.B. Ltd (Birkenhead)
33329 grt (22335 nrt) 57372 dwt
210.01 [202.50] x 32.25 x 16.41 m
(689.0 [664.8] x 105.8 x 53.8 ft)
2 single-acting 6-cyl. Sulzer Diesel engines
17400 BHP;15 kn
by G. Clark & N.E.M. Ltd
(Wallsend-upon-Tyne)

1976 September 8th: launched by Lady Verdon-Smith, the wife of the chairman of Lloyd Bank International, Sir Reginald Verdon-Smith; first standard petroleum products tanker type "StaT 55" built by Messrs Cammell Laird and first of five sisters ordered by the Vlasov Group.
1977 May 2nd: completed and delivered to Algol Shipping Co. Ltd, London; British flag.
1978: transferred to S.S.I. NAV I Ltd (Bermuda).
1979: transferred to Alkor Shipping Co. Ltd, London; Silver Line management.
1982 July 10th: severe fire damage while 130

The second Algol being ready for launch on 8th September 1976 and (below) the flag-changing ceremony on the occasion of her delivery on the following 2nd May.

Tankers, ore/oil and product carriers

The *Alvega*, under the command of Capt. A. G. Tester, was employed during the Falklands war as a support ship to the British South Atlantic Fleet; in this 16th August 1982 photograph taken off Ascension Island she is seen linked to *RFA Brambleleaf* and pumping over fuel oil.
(Matteo Parodi, Monaco)

1993: re-named *Lundina*.
1995: sold to Grecian Shipping Co. and re-named *Santa Lucia*; Navitankers Management Inc.; p.o.r. La Valletta, Malta.
1998: in service.

ALVEGA II - WHITE SEA - CAM KOLE (1977)

Tanker
Builders: Cammell Laird S.B. Ltd (Birkenhead)
33329 grt (22335 nrt) 57372 dwt
210.01 {202.50} x 32.25 x 16.41 m
(689.0 {664.4} x 105.8 x 53.8 ft)
1 single-acting 6-cyl. Sulzer Diesel engine
17400 BHP;15 kn
by G. Clark & N.E.M. Ltd
(Wallsend-upon-Tyne)

miles off Cristobal, en-route from Guaymas to Pajaritos; ten casualties; two days later arrived at Cristobal to be repaired.
1983: transferred to Ocelot Inc., Monrovia; re-entered service with the new name of *Ocelot*, Liberian flag. Shipping Management S.A.M., Monaco Monte Carlo.
1985: sold to Jaspidea Shipping Corp., Manila; renamed *Dagli*, Filipino flag.
1989: sold to Iver Bugge A/S, Oslo and re-named *Iver Lundina*.

1974: ordered with the intended name of *Alkaid*.
1977 March 8th: launched for Alvega Shipping Co. Ltd, London with the name of *Alvega* by Mrs Van Pelt (wife of the Senior Vice President of Citybank N. A.), as in the meanwhile the tanker bearing this name was sold for scrap; British flag.
1978: transferred to S.S.I. NAV II Ltd (Bermuda).
1979: transferred to Finance for Shipping Ltd, London; Silver Line management.
1982: called up by the British Ministry of Defence and employed as supply vessel during the

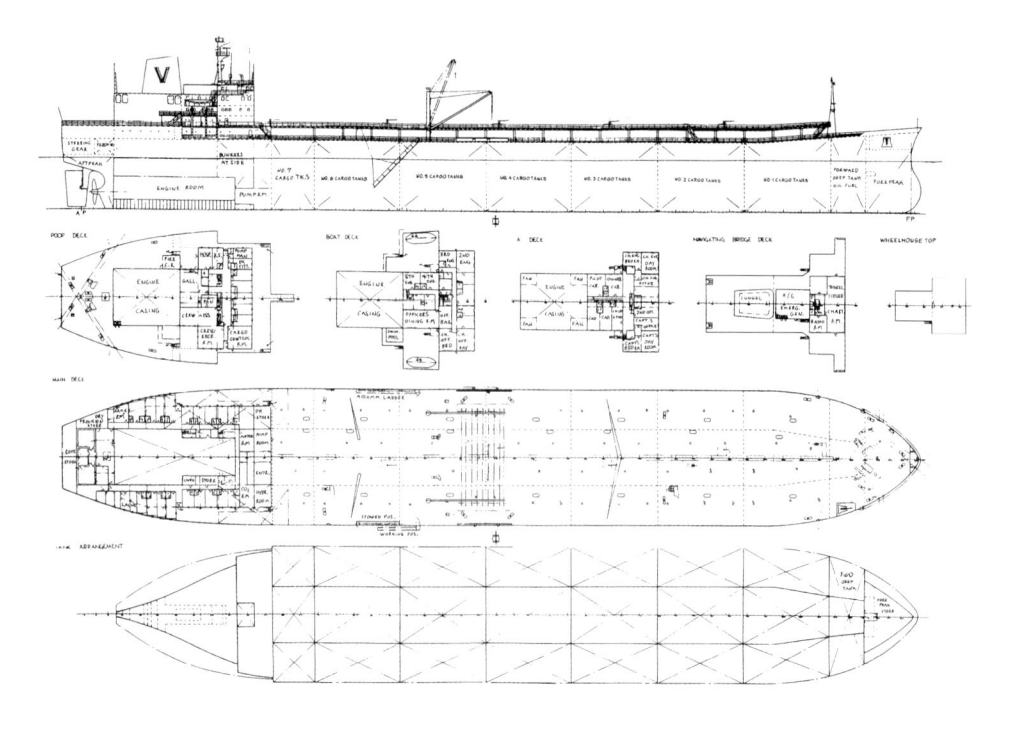

The general arragement plan for the succesful Cammell Laird's standard "StaT 55" tanker, of which Vlasov ordered five in the mid-'seventies, christened *Algol*, *Alvega*, *Alice Redfield*, *Almak* and *Alvenus*.

Tankers, ore/oil and product carriers

Falklands War.
1983: transferred to Investors in Industry PLC, London; Silver Line management.
1986: transferred to the Gibraltar flag with the name of *White Sea*.
1996: transferred to Vlasov Shipholdings Inc., Liberia; Shipping Management S.A.M.; re-named *Cam Kole*; British flag.
1998: in service, last vessel of the old Vlasov fleet still owned by the Group.

ALNAIR II (1977-1983)

Tanker (formerly *Iddi*, then *Miztli*)
Builders A/B Gotaverken Arendal (Gothenburg)
36788 grt (23615 nrt) 74480 dwt
239.28 [231.65] x 36.89 x 16.46 m
(785.0 [760.0] x 121.0 x 54.0 ft)
1 single-acting 9-cyl. Diesel engine
16300 SHP; 14.5 kn
by builders

1965: delivered to Skibs A/S Motortank & A/S Oljefart II, Oslo, and entered service as *Iddi*.
1977: bought by Vlasov Group subsidiary Alnair Shipping Corp., Monrovia and re-named *Alnair*; Liberian flag.
1982: sold to Cia. Naviera Mexicana Santa Andres S.S. de C.V., Mexico, and re-named *Mitzli*.
1983 July 13th: arrived at Kaoshiung to be broken up.

ALMALAZ (1978-1986)

Oil and Chemical Tanker
(formerly *Birgitta Fernström*, *Saga Surf*, *Osco Surf*, *Serra Trader*, then *Penelope*)
Builders: Oresundsvarvet A/B (Landskrona)
25503 grt (20313 nrt) 50875 dwt
221.14 [210.07] x 30.79 x 9.00 m
(725.5 [689.2] x 101.0 x 29.6 ft)
1 single-acting 8-cyl. Diesel engine
16800 BHP; 15 kn
by A/B Gotaverken (Gothenburg)

1965: completed and delivered as *Birgitta Fernström* to A.K. Fernström Rederier, Karlshamn.
1972: sold to I/S Ole Schroder A/S and re-named *Saga Surf*.
1975: laid up, re-named *Osco Surf* and put up for sale.
Sold to Serra Shipping Ltd and re-named *Serra Trader*.
1978: sold to the Vlasov Group subsidiary

The 1977-built *Alvega* is still in service for the Vlasov Group as the *Cam Kole*.

In 1980 the *Almalaz* was transferred to Vlasov's subsidiary Amar Line and her hull was re-painted in the distinctive green colour adopted by the Saudi Arabian company.
(Arnold Brereton, Monaco)

Andweel Holdings Inc. Monrovia and re-named *Almalaz*; Liberian flag.
1980: transferred to Amar Line subsidiary Najd Maritime Transport Co. Ltd; p.o.r. Jeddah.
1986: sold to Horafa Shipping Co. Ltd, Monrovia; re-named *Penelope*.
1998: in service.

ALICE REDFIELD - ALKES III - SALINA (1978- 1986)

Tanker
(then *Dagitab, Dagfred, Iver Christina, Skaw Princess, Mare Princess*)
Builders: Cammell Laird S.B. Ltd (Birkenhead)
33329 grt (22335 nrt) 57372 dwt
210.01 [202.50] x 32.25 x 16.41 m
(689.0 [664.4] x 105.8 x 53.8 ft)
1 single-acting 6-cyl. Sulzer Diesel engine;
17400 BSHP; 15 kn
by G. Clark & N.E.M. Ltd
(Wallsend-upon-Tyne)

1978: intended to be launched as the *Alkor*, during her building the owners signed a long time charter with the American Atlantic Richfield Corp. which chose for her the name *Alice Redfield*. Delivered to Vlasov Group subsidiary S.S.I. Nav III Ltd (Bermuda); British flag. September 9th: she broke a 1959 world-record discharging 34,914 barrels of crude oil per hour.
1979 January 4th: she improved her own record discharging 363,786 barrels of crude oil in exactly ten hours.
1980: transferred to Alkes Shipping Co. Ltd, London, and re-named *Alkes*.
1983: transferred to Salina Finance Corp., Monrovia; re-named *Salina*; Liberian flag; Shipping Management S.A.M.
1986: sold to Buddleia Shipping Corp. Manila and re-named *Dagitab*; p.o.r. Manila.
1988: re-named *Dagfred*; p.o.r. Oslo; same owner; J.P.P. Shipping Corp. Oslo managers.
1989: sold to Iver Bugge Corsair Holding A/S, Larvik, and re-named *Iver Christina*.
1991: sold to Corsair Holding Inc. (Wind Management A/S, Oslo) and re-named *Skaw Princess*.
1992: sold to Mare Princess Inc. Monrovia, subsidiary of Mare Maritime Co. Piraeus, and re-named *Mare Princess*.
1998: in service.

ALMAK II - ALVA SEA II (1978- 1994)

Tanker (then *Provence*)
Builders: Cammell Laird S.B. Ltd (Birkenhead)
33329 grt (22335 nrt) 57372 dwt
210.01 [202.50] x 32.25 x 16.41 m
(689.0 [664.4] x 105.8 x 53.8 ft)
1 single-acting 6-cyl. Sulzer Diesel engine
17400 BHP; 15 kn
by G. Clark & N.E.M. Ltd
(Wallsend-upon-Tyne)

1978 March 10th: launched. October: completed and delivered to Lloyds Leasing Ltd, London;
1983: owner unchanged but management transferred to Shipping Management S.A.M., Monaco Monte Carlo.
1994: re-named *Alva Sea* and put up for sale; sold to Provence Shipping Co.; re-named *Provence*; Arminter SAM, Monte Carlo managers; p.o.r. Valletta, Malta.
1998: in service.

KWAI (1978 - 1982)

Tanker
(formerly *Axel Brostrom*, then *Catemaco*)
Builders: A/B Gotaverken (Gothenburg)
42360 grt (24350nrt) 74480dwt
237.74 [231.65] x 36.91 x 12.71 m
(780.0 [760.0] x 121.1 x 41.7 ft)
1 Gotaverken Diesel engine
16300 SHP; 14.5 kn
by builders

1965: completed and delivered as *Axel Brostrom* to Brostrom Group, Gothenburg.
1978: sold to Vlasov and re-named *Kwai*; registered in Hong Kong for Triesto Shipping Enterprise.
1981 October 12th: while in Algeciras she was hit by the container ship *Saint Louis*. The bulbous bow of the latter vessel strongly penetrated the *Kwai*'s starboard side, opening a 15 sqm underwater breach in the tank no. 1, which was fortunately empty and de-gassed. After a provisional patch was bolted in Gibraltar, the *Kwai* resumed her passage to Terneuzen, Netherlands where her cargo of virgin nafta was unloaded. Later repaired in Cadiz.
1982: sold to Santa Eugenia Naviera, Monzanillo (Mexico), delivered to her new owner in New Orleans and re-named *Catemaco*.
1983 June 16th: arrived at Kaoshiung to be broken up by Kim Tai Steel Enterprise.

Tankers, ore/oil and product carriers

ALVENUS II (1979- 1991)

Tanker
(then *Alifax, Crimson King, Plate Princess*)
Builders: Cammell Laird S.B. Ltd (Birkenhead)
33329 grt (22335 nrt) 57372 dwt
210.01 [202.50] x 32.25 x 16.41 m
(689.0 [664.4] x 105.8 x 53.8 ft)
2 single-acting 6-cyl. Sulzer Diesel engines
17400 BHP;15 kn
by G. Clark & N.E.M. Ltd
(Wallsend-upon-Tyne)

1979 March 3rd: completed and delivered to Lloyds Leasing Ltd, London; last of five product carriers built by Messrs Cammell Laird for the Vlasov Group; Shipping Management S.A.M, Monaco.
1983: owner unchanged but management transferred to Silver Line.
1984 July 31st: aground in Calcasien Channel, 40 miles South of her Lake Charles destination, with a full load of Venezuelan crude; the hull was buckled for 50 metres from the stem, so that the forefoot was three metres over the keel line; with special reinforcements to her bent hull the vessel sailed for the Swedish Cityvarvet shipyard where she underwent major repairs.
At the time it was reported as one of the most spectacular "surgery operations" ever carried out on a vessel.
1988: re-named *Alifax*; V.Ships management.
1991: sold to Ocean Marine S.A. Malta and re-named *Crimson King*; p.o.r. Valletta.
1996: transferred to Ocean Marine S.A. Buenos Aires and re-named *Plate Princess*.
1998: in service.

ALTANIN II (1980- 1980)

Product carrier (formerly *Athelqueen*)
Builders: Davie Shipbuilding Ltd (Lauzon)
24132 grt (16297 nrt) 39728 dwt
182.88 [173.74] x 32.31 x 15.17 m
(600.0 [570.0] x 106.0 x 49.8 ft)
1 single-acting 7-cyl. Sulzer Diesel engine
14000 BHP;15.25 kn
by G. Clark & N.E.M. Ltd
(Wallsend-upon-Tyne)

1977: completed and delivered as *Athelqueen* to Athelstane Tankers Co. Ltd, London.
1978 August 21st: while in Cadiz for repairs an explosion in a tank killed three men. October 20th: collided with S/S *Maipo* while two and a half miles south of the Verrazano-Narrows Bridge; subsequently repaired.
1980: bought by Altanin Tanker Co. Ltd, London, a subsidiary of Dene Shipping Co. Ltd, controlled by Navigation & Coal Trade Ltd and owned by the Vlasov Group; re-named *Altanin*; British flag. Silver Line management. September 2nd: cleared Khorranshahr with a full load of comestible oil; on the 23rd she remained trapped off Jazirat-Abu Davo (Shat-al-Arab) owing to the Iran-Iraq war.
1982: abandoned to the insurers as war loss.

ALRAI II - BARAKA - LUCERNA (1980-1986)

Product carrier
(formerly *Athelmonarch*, then *Dakila, Dagrun, Quebec, Silver Shing, Yong Chi*)
Builders: Davie Shipbuilding (Lauzon)
24132 grt (16297 nrt) 39728 dwt
182.88 [173.74] x 32.28 x 15.17 m
(600.0 [568.7] x 105.9 x 49.8 ft)
1 single-acting 7-cyl. Sulzer Diesel engine
14000 BHP;15.25 kn
by G. Clark & N.E.M. Ltd
(Wallsend-upon-Tyne)

1977 May: completed and delivered as *Athelmonarch* to Athelstane Tankers Co. Ltd, London.
1980: bought by Alrai Shipping Co. Ltd, London, a subsidiary of Dene Shipping Co. Ltd, controlled by Navigation & Coal Trade Ltd and owned by the Vlasov Group; re-named *Alrai*; British flag.
1982: transferred to Divichi Navigation Co., Monrovia, re-named *Baraka*, Liberian flag.
1983: re-named *Lucerna*, same owner; Shipping Management S.A.M. Monaco Monte Carlo.
1986: sold to Mauve Queen Shipping Co., Manila, and re-named *Dakila*; Filipino flag.
1988: sold to J.P.P. Shipping A/S Oslo and re-named *Dagrun*; p.o.r. Oslo.
1989: sold to Quebec Shipping Line and re-named *Quebec*; management entrusted to V.Ships.
1996: laid up, re-named *Silver Shing* and put up for sale; bought by Guangzhou Maritime Transport Group Co. of Guangzhou, China and re-named *Yong Chi*.
1998: in service.

TANKERS, ORE/OIL AND PRODUCT CARRIERS

On 5th May 1984, the *Al Ahood*, while sailing off Kharg Island, was hit by an Iraqi missile at the level of the stern upperworks which were completely burnt out.
(Alberto Bisagno, Genoa)

AL AHOOD (1980-1984)
Tanker
(formerly *Conoco Britannia*, *Venture Britannia*)
Builders: Astilleros Españoles S.A. (Cadiz)
58277 grt (47609 nrt) 117710 dwt
279.30 [258.50] x 42.00 x 20.02 m
(916.3 [848.1] x 137.8 x 65.7)
1 single-acting 7-cyl. Sulzer Diesel engine;
23200 BHP;16 kn
by builders' Manises Works (Valencia)

1972: completed and delivered to Worldwide Transport Inc., Monrovia as *Conoco Britannia*, Liberian flag.
1978: re-named *Venture Britannia*; same owner.
1980: bought by Vlasov Group subsidiary Alioth Shipping Corp., Monrovia and re-named *Al Ahood*.
1981: transferred to Amar Line, Jeddah; Saudi Arabian flag.
1984 May 5th: hit by an Iraqi missile at 28.07N, 51.06E, 80 miles off Kharg Island. On fire, abandoned by crew; later towed and anchored 30 miles off Bahrain harbour. September 12th: delivered at Kaoshiung to First Copper & Iron Industrial Co. Ltd to be broken up.

AL KHLOOD (1980 - 1990)

Chemical tanker (formerly *Silverpelerin*, then *Arctic Tar*, *Arctic Star*)
Builders: Krögerwerft GmbH (Rendsburg)
4599 grt (1893 nrt) 6720 dwt
112.80 [104.00] x 16.50 x 8.90
(370.1 [341.2] x 54.1 x 29.2 ft)
1 single-acting 4-stroke 12-cyl. Diesel engine
5644 SHP;14 kn
by Mirrlees (Stockport)

1972 December 22nd: launched with the name *Silverpelerin*.
1973 April: completed and delivered to Silver Tankers Ltd, a subsidiary of Silver Line, London.
1980: transferred to the Amar Line, Jeddah and renamed *Al Khlood*, p.o.r. Jeddah.
1990: sold to Athenea Partnership Shipping, renamed *Arctic Tar*, St. Vincent & Grenadines flag, ICTB, Belgium managers.
1995: re-named *Arctic Star*, same owner.
1998: in service.

Silver Line

In 1973 the Vlasov Group bought a 50% share of Shipping Industrial Holdings Ltd and four years later took full control of this company, one of the largest British shipping concern. Thanks to this brilliant move Silver Line Ltd, a subsidiary of Shipping Industrial Holdings Ltd, became the property of Vlasov Group. This instantaneously brought to Vlasov Group 21 modern vessels plus two 120,000 dwt Ore/Oil carriers under construction in Japan.

Silver Line Ltd, nowadays the commercial arm of the Vlasov Group, after its 1977 merging with the 1937 Vlasov-founded Navigation & Coal Trade Ltd, occupies an important place in the history of British Shipping; founded with the name Silver Line in 1925, its origins can be traced back to 1908, when its founder, Stanley Miller Thompson, opened St. Helen's S.S. Co. Ltd, starting operations with the 4,800 dwt freighter *Silverbirch*. Although the long history of the Silver Line does not come within the scope of this book, it seemed appropriate to reproduce in the following pages the booklet "Half a Century of Silver Line", published in 1975 to commemorate the company's anniversary and in occasion of the Vlasov take-over. Further information on the present activity of Silver Line Ltd can be found in the chapter "V.Ships today and tomorrow".

The medal struck in 1975 by Silver Line to commemorate its 50th anniversary.

Silver Line

Half a Century of Silver Line

Silver Line Limited, which today is responsible for some 2,000,000 tons d.w. of shipping, was registered as a public company on 24 November, 1925. In a sense, however, the first chapter in its 50-year history was written several years earlier, in 1908, when Stanley Miller Thompson, until then engaged as a London shipbroker and manager, went into partnership with his brother John as managing owners of the St. Helen's Steamshipping Co. Ltd. The latter owned one vessel, the steamship "Wearmouth", and on the change of ownership her name became "Silverbirch", thus giving birth to a prefix which has been perpetuated by nearly 70 Silver Line vessels since that date.

The Thompson family, part of the famous Sunderland shipbuilding family, were to remain in control of Silver Line for some 20 years, then to be succeeded by another North East coast family, the Barracloughs, whose influence was to last for the next 30 years — until 1974, in fact. The change had a significant effect on the nature of Silver Line's operations, for prior to 1951, during the time of the Thompsons, it functioned exclusively as a cargo liner company, operating services in all parts of the world; from that date onward, however, it has been involved entirely in the tramp trades and timecharter market.

Widely different though the philosophies of the two families undoubtedly were, there was some common ground. For example, throughout its history, Silver Line has been involved in many joint ventures with other companies, in many cases acting as the operating company, in recognition of its expertise and experience in owning and managing ships. Another characteristic of the Company's operations is that its ships have always earned most of their revenues in foreign waters, the majority of them seldom visiting UK ports. Moreover, since its inception, Silver Line has been an innovator in ship design. It was one of the pioneers of the marine diesel, the original vessels in the fleet being among the first ships to have Doxford engines. "Silverlaurel", which entered service in 1939, was the first of her size and class to be powered by steam turbines. The "Silverbriar" class of '48, with their twin funnels (the forward of which housed the chartroom, wheelhouse and master's accommodation) and transatlantic liner styling, helped to set new standards in cargo ship design. "Chelsea Bridge" (105,000 tons d.w.) made history when she was commissioned in 1967 as the largest bulk carrier under British flag. Moreover, the chemical tankers "Silverfalcon" and "Silvermerlin" were the first British ships to be built with stainless steel tanks, enabling them to carry a wide range of bulk chemical cargoes — a concept carried forward on a bigger scale in "Silverosprey" and "Silvereagle".

Such developments are among the more important landmarks in the Silver Line history which this booklet seeks to outline, landmarks like the order for six ships to be built on the Wear in 1926/27, the salvation of Sunderland shipbuilding in those difficult years; the war years, when no fewer than eleven of the fleet of 18 were lost; the acquisition of the company by Dene in the late '50's; the establishment of Seabridge Shipping Ltd. in 1965, one of the largest bulk carrier consortia in the world, with Silver Line as a founder member; the amalgamation with Shipping Industrial Holdings in 1971; and the subsequent acquisition of SIH by Navcot Shipping (Holdings) Ltd. and Silver Line's appointment to operate all UK tonnage for the Vlasov Group.

These events have brought Silver Line into the second half of its personal century with a fleet substantially larger than at any other period in its history, forming part of one of the largest in international shipping. It is a role for which it has been well prepared.

4. Stanley Miller Thompson, founder Chairman of Silver Line, who steered the Company through the first two decades of its history.

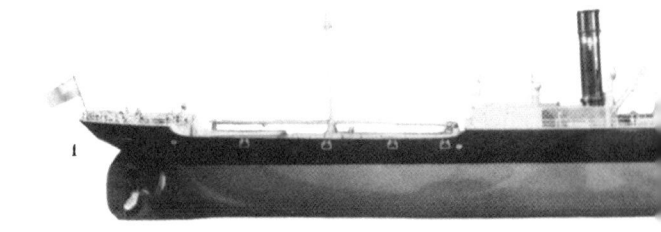

1. The steel screw steamer "Silverbirch", 4,800 tons d.w., the first 'silver' ship. Built in 1905 as the "Wearmouth" by Joseph L. Thompson for St. Helen's Steamshipping Co. at a cost of £42,500. Disposed of before the actual formation of Silver Line.

2. One of the first of many Silver Line launches, that of the "Silveray" in 1925 at Joseph L. Thompson's famous North Sands yard.

3. Members of the "Silveray" launch party on 25th April, 1925
Left to right: Major R. N. Thompson, Col. J. Lynn Marr, Mrs. J. Lynn Marr, Sir James Marr, Bart, C.B.E., J.P., Mr. Charles D. Doxford, Mrs. Charles Doxford, Lady Marr, Mr. J. W. Thompson, Mr. S. M. Thompson, Mr. Charles Doxford, Mrs. Robert Thompson, Mrs. R. N. Thompson, Mrs. W. B. Marr, Capt. T. A. Ensor, Mrs. H. C. Coatsworth, Mrs. K. O. Keller, Mr. John Catto, Mr. K. O. Keller, Mrs. John Catto, and Mrs. T. A. Ensor.

SILVER LINE

5. "Silverash", the first of the six ships built on the Wear during 1926/27, photographed on her commissioning. The contract price was £190,000 per ship and the order proved to be the salvation of the Sunderland shipbuilding industry.

6. One of the largest ships in the current Silver Line fleet, the 142,600 tons d.w. OBO (oil/bulk/ore) carrier "Silver Bridge". She is one of seven Silver Line ships in service with the Seabridge bulk carrier consortium.

7. Capt. W. J. Irvine, master of the "Silverbeech", who made an interesting contribution to Silver Line history in 1935

8. "Silverelm", built in 1924, was one of the first ships to be powered by Doxford diesel engines.

9. Services operated by the Company during the cargo liner years included:

Outward: U.S. and Canada, Atlantic and Pacific, to Philippines, Shanghai and Hong Kong. U.S. Atlantic to Mediterranean, Egypt, Red Sea, Persian Gulf, Ceylon, India, Pakistan and Burma. Gulf of Mexico to South and East Africa. British Columbia and U.S Pacific to South and East Africa. British Columbia and U.S. Pacific to Philippines, N.E.I., Malaya, India, Pakistan, Burma, Ceylon and Persian Gulf.

Inward: Philippines, N.E.I., Malaya and Ceylon to Eastern Canada and U.S. Atlantic. West Coast India, Pakistan and Colombo to Eastern Canada and U.S. Atlantic. Persian Gulf, Ceylon, Pakistan, India, Burma, Malaya, N.E.I., and Philippines to U.S. Pacific and British Columbia.

10. Henry Barraclough, who took over the helm from Stanley Thompson, and who was Chairman from 1947 until 1959, being succeeded first by his brother Willie, 1959-66 and later by his son David, 1968-74.

Half a Century of Silver Line

1925 Silver Line Ltd. registered as public company, 24 November, with headquarters at 80 Bishopsgate, London EC2. Formed to acquire six vessels owned by four private companies — St. Helen's Steamshipping Co. (1912) Ltd., Mount Steamship Co. Ltd., Silvercedar Shipping Co. Ltd., and Way Shipping Co. Ltd. Stanley Miller Thompson elected Chairman. Stanley and John Thompson Ltd. appointed managers and Kerr Steamship Co. Inc., loading brokers. Vessels already operating on Kerr Line berth from U.S. Atlantic and Gulf of Mexico to Far East.

1926/7 Orders placed for six vessels of "Silverash" class, 8,900 tons d.w., three with Joseph L. Thompson & Sons Ltd., three with Sir James Laing & Sons Ltd. (then the oldest surviving firm on River Wear and builders in 1875 of famous clipper ship, Joseph Conrad's "Torrens"). Vessels subsequently enter service on New York-Far East-New York service — the 'Round World Service'.

1928 Original six vessels transferred to pioneer new service between Pacific Coast and Orient, thus providing first regular services between Pacific Coast and many parts of the East.

1929 Orders placed with Thompsons (three) and Harland and Wolff Ltd., Belfast (four) for seven twin-screw motor vessels of "Silverpalm" class. Total cost — £1,600,000.

New, fast service from American Pacific coast to Calcutta, the 'Pacific India' service, inaugurated.

1930 Partnership agreement with powerful Dutch combine, Java-Bengal Line, results in new joint 'Silver-Java-Pacific Line' service, each company allocating seven vessels.

Five smaller Silver Line vessels transferred to inaugurate 'States-India-Persian Gulf' service, from Gulf of Mexico to west coast of India and Persian Gulf.

1932 Important new service opens in 'Silver-Java-Pacific Line' — U.S. west coast via Panama to Gulf of Mexico ports, South Africa, India, Straits Settlement, Dutch East Indies, Philippines, and across Pacific to U.S. west coast; service later extended to cover Canadian west coast ports.

Agreement with Prince Line to integrate sailings on 'Round World Service' into common joint schedule.

1933 Non-Japanese lines precluded from carrying cargo to Japanese ports; 'Round-World-Service' revised accordingly.

1934 Demand for cargo space from Gulf of Mexico to South Africa warrants additional monthly service to African ports.

Accommodation provided for 12 passengers in certain vessels (a new innovation).

1935 Twin daughters born to passenger aboard "Silverbeech", safely delivered by Capt. W. J. Irvine (a bachelor); girls later christened Sylvia Beech and Silver Beth.

Increased cargo movement from East to America justifies augmenting 'Round-World-Service'. New, expanded service called 'States-Africa-East Indies' service.

1936 Silver Line vessels forced to continue proceeding via Cape as result of Spanish Civil War.

1937 'Silver-Java-Pacific Line' transpacific service extended to include Persian Gulf.

Agreement with Leif Hoegh, Oslo, for latter to provide vessels as required to cover Silver Line's expanding commitments in transpacific trades.

1938 Comprehensive pension scheme introduced for seagoing and office personnel (a lead subsequently followed by many other companies).

1939 Two new services inaugurated — U.S. Atlantic and Gulf of Mexico ports to Persian Gulf and India, called 'American-Persian Gulf' and 'U.S./India Line' respectively.

Outbreak of World War II; section of Head Office transferred to Over Dinsdale Hall, Teesside, as precautionary measure.

1940 Silver Line moves into new HQ — Palmerston House, Bishopsgate, opposite original Head Office. Over Dinsdale Hall staff return permanently to London.

Fleet, now numbering 18 ships, requisitioned; vessels used for transportation of aeroplanes, tanks and Army supplies.

11

12

11. "Silverlaurel", which entered service in 1939, was the first vessel of her type and size to have steam turbines instead of diesel engines. Pictured in wartime colours in Liverpool. She was sunk off Plymouth in 1944.

12. Another of the eleven wartime losses, "Silverwillow", built in 1930 and sunk off Spanish coast, 1942.

Company berths, become base for large numbers of requisitioned ships.

Silver Line's Technical Adviser, R. Cyril Thompson, nephew of Stanley and Managing Director of Joseph L. Thompson, jointly heads Admiralty Merchant Shipbuilding Mission to U.S.A., taking plans of standard Thompson trampship 'Empire Liberty', prototype of over 2,700 'Liberty' ships built in U.S.

First wartime casualty, "Silverpine", sunk in N. Atlantic with master, seven officers, 25 crew.

1941-4 Ten more ships sunk in action (details — inside back cover), wartime casualties totalling 11 vessels, 104 British officers, seamen and D.E.M.S. gunners, and 97 Chinese seamen.

SILVER LINE

1944 Delivery of "Silveroak", first wartime replacement and largest Silver Line vessel so far built — 10,376 tons d.w.

1946 Berth rights (in trades previously served by Company vessels) of Kerr Steamship Co. Inc. acquired, together with share capital of Kerr-Silver Lines (Canada) Ltd.

Two American-built 'Liberty' and four Canadian 'Victory' ships added to fleet.

1947 Silver Line, hitherto managed by Stanley and John Thompson Ltd., becomes self-managed company. Stanley & John Thompson (Management) Ltd., Henry Barraclough, member of Board since 1937, and also director of Dene Shipping Ltd., becomes Chairman of Silver Line.

1948 Delivery of distinctively-designed "Silverbriar" and "Silverplane", first ships in post-war newbuilding programme.

Silver Line acquire Stanley & John Thompson (Management) Ltd.

15. "Silveroak", built in 1944 as first wartime replacement vessel, bridged pre-war "Silverlaurel" and post-war "Silverbriar" designs. She was powered by a twin-screw Doxford diesel, it being impossible in wartime to obtain single-screw propelling machinery to give required power for a vessel of her size (16,500 tons displacement). She was, in fact, the largest Silver Line vessel built up to that date and had accommodation for 12 passengers on a separate passenger deck and (under special regulations then obtaining) comfortably accommodated up to 48 on occasions.

13/14. The remarkable transatlantic liner styling of "Silverbriar", 10,700 tons d.w., incorporated a false 'funnel' forward, containing captain's accommodation, wheelhouse, chartroom, wireless room, standard compass and radar. Delivered, 1948, to inaugurate the then merged 'American-Persian Gulf' and 'U.S./India Line' service.

1951 Silver Line alters course, abandoning liner trades and switching to tramp trading and time-charter.

Acquisition of "Riodene" and re-naming to "Silvertarn"; departure from choice of tree names to that of geographical features for ships' names reflecting Barraclough family interest in Lake District.

1952-3 The Company's first oil tankers constructed, "Silverdale", 16,700 tons d.w., by Lithgows, "Silverbrook", 16,630 tons d.w., by Smith's Dock.

1956 Delivery of 10,000 tons d.w. m.v. "Silverdene" (a name that was to have particular significance the following year).

1957 Silver Line becomes fully-owned subsidiary of Dene Shipping Co. Ltd., cementing association of many years resulting from Henry Barraclough's close interest in both.

Agreement with Cardiff Ship Store Company, a Dene subsidiary, to supply ships in Silver Line fleet (which it continues to do to this day).

1958 First of the ore carriers, "Silversand" and "Silvercrag", delivered to St. Helen's Shipping Co. Ltd., consortium comprising Silver Line, John I. Jacobs, Joseph L. Thompson and Sir James Laing.

16. "Silverbeach" loading grain in Ohio. The former "Totem Star" (ex "Norse Coral") was one of the first ships to be equipped with portable car decks, fitting her for the interesting trading pattern she was to develop with her sister ship "Silversea", involving European and Japanese cars and North American grain.

17. "Totem Queen", owned by Fulcrum Shipping Co. Ltd., of Nassau and managed by Silver Line, before being bought by the latter in 1964 and re-named "Silversea". In April of that year she had the distinction of being the first vessel to open the St. Lawrence Seaway on the earliest resumption of shipping in the Seaway's history.

18/19. The ship that was to become the first bulk carrier in the Seabridge fleet seen fitting-out at the yard of Sir James Laing (18). She was launched "Silverhow" (a 'how' being a low hill — a typical Lake District feature) and re-named "Tower Bridge" (19).

1960 Two further ore carriers, "Aldersgate" and "Bishopsgate", delivered, this consortium (Silver Line, Jacobs, Thompsons and Laings joined by British Steel Corporation). Delivery of "Silverisle" to Silver Isle Navigation (Bermuda) Ltd., a consortium of Silver Line and Jacobs.

1963 First venture into car carriers; Silver Line appointed managers of Totem Line, owners of "Totem Star" (ex "Norse Coral") and "Totem Queen" (ex "Norse Reef"), two car and bulk carriers, subsequently bought by Silver Line and renamed "Silverbeach" and "Silversea" respectively. Trading pattern: German cars to Canada and U.S.A., grain back from Great Lakes to Europe. Developed at later date into: U.K. cars to U.S.; U.S. grain to Japan; Japanese cars to U.S. Gulf; and U.S. grain back to U.K./Continent — a five month round trip.

1964 Order placed for (then very large) bulk carrier, 34,400 tons d.w., to replace three comparatively small and slow vessels, "Silverfell", "Silverlake" and "Silverforce"; launched following year as "Silverhow", later re-named "Tower Bridge".

First issue of Silver Line Monthly News Sheet published (now Silver Line *Newsletter*). Issue No. 3 reports first consignment of new cargo — 2,000 tons of bagged mud from U.S. Gulf to Benghazi; quotes master, Capt. G. F. Chivers: "Now I've carried everything . . .!"

Introduction of scheme for Head Office staff to make short 'familiarisation' voyages in Company ships.

1965 Formation of Cornhill Shipping Co. Ltd., wholly-owned subsidiary, as owners of new chemical tanker "Silverkestrel" for carriage of liquid caustic soda from the Mersey to Carrickfergus on contract with Courtaulds Ltd.

Seabridge Shipping Limited formed by Silver Line, Bibby Line, Britain Steamship and H. Clarkson to operate large bulk carriers as single fleet. Consortium later joined by C. T. Bowring and subsequently by Houlder Bros. and Huntings. Silver Line director, Derek Hall, appointed to manage new Company, based at Palmerston House. Silver Line's "Tower Bridge" first vessel to enter service with Seabridge.

21. One of the first two ore carriers, "Silvercrag", 15,500 tons d.w., delivered in 1958. Her funnel bears the old-style markings, to which the current Silver Line symbol, introduced in 1967, owes its origins.

20. "Silverkestrel" on her commissioning in 1965. Designed for the carriage of caustic soda, she was the first Silver Line ship to be ordered from a foreign shipyard — Ekeroths of Norrkoping, Sweden. After ten years' trading, during which period she was 'jumboised' to give her an increased deadweight, she was sold earlier this year.

1966 Silver Line's chemical tanker-owning subsidiary, Cornhill Shipping, takes delivery of "Silverfalcon", first British ship fitted with stainless steel tanks for carriage of acids and other chemicals in bulk.

1967 "Silvercove" joins fleet, one of two 18,650 tons d.w. bulk carriers specially designed for carriage of British Columbian packaged lumber; first Silver Line ship built in Japan, and first 'foreign' ship built by Namura Shipbuilding, Osaka.

Silver Line takes delivery of 105,000 tons d.w. "Sigsilver", later re-named "Chelsea Bridge", for operation in Seabridge.

Contract placed with Cammell Laird for one 9,000 tons d.w. chemical tanker at cost of £1,900,000 — exactly ten times cost of "Silverash" class of similar size (though different type) 40 years earlier.

Name of Cornhill Shipping changed to Silver Chemical Tankers Ltd.

1968 General Chemical Chartering Ltd. formed as broking company in partnership with ICI Pension Fund (wholly-owned by latter as of earlier this year).

1969 Bishopsgate Shipping and Silver Isle Navigation become 100% subsidiaries; interests in St. Helen's Shipping disposed of.

G. P. Seamen manning concept introduced, "Silvereid"; later extended to other vessels in fleet.

1970 Chemical tankers, "Silverosprey", "Silverharrier" and "Silvereagle", each of over 6,000 tons d.w., delivered.

"Silverharrier", with "Silverhawk", 10,400 tons d.w., chartered to Australian interests for carriage of sulphuric acid to I.C.I.A.N.Z. fertilizer plants.

1971 Dene Shipping amalgamated with Shipping Industrial Holdings Ltd.; combined fleet of ships in service and on order totals 1,500,000 tons d.w.

1972 Dene Shipping assumes responsibility for shipowning activities of S.I.H.; day-to-day administration and overall operational control undertaken by Silver Line.

Four large newbuildings delivered for service with Seabridge — "Silver Bridge" and "Eden Bridge", both 142,000 tons d.w., "Severn Bridge" and "Stirling Bridge", both 118,000 tons d.w.

1973 Latest newbuildings — bulk carrier, "Silverdon", and chemical tanker, "Silverpelerin" — enter service.

Orders placed for two 120,000 tons d.w. OBO carriers with Mitsubishi at total cost of £20,000,000; among first OBO's on order to comply with new I.M.C.O. requirements for increased safety at sea.

Acquisition of Seabridge Shipping Services Ltd., the managing company of the Seabridge consortium.

1974 S.I.H., incorporating Dene Shipping and Silver Line, acquired by Navcot Shipping (Holdings) Ltd.

22/23. Equipped with stainless steel tanks for the carriage of a wide range of chemical cargoes, "Silvermerlin" was delivered in 1968; now serving in recently-formed Interchem chemical tanker pool.

24. "Silverdon", named after the Scottish River Don, and one of the latest bulk carriers to join the Silver Line fleet. A dry bulk carrier of 32,300 tons d.w., she is specially strengthened for the carriage of ore.

25. Silver Line's most modern chemical tanker, and one of the most up-to-date of her type in the world, "Silverpelerin". Named, like the other vessels in the chemical tanker fleet, after a bird of prey — the peregrine falcon. The French (pelerin) spelling was chosen since the ship was designed to trade mainly between Canada and the U.S.A.

Silver Line appointed operating company for Vlasov Group's British-flag tonnage, comprising two tankers and two ore/oil carriers with five product carriers on order, a total fleet in service and on order of more than 1,000,000 tons d.w.

1975 Group's Head Office transferred from Bishopsgate (after nearly 50 years) to 43 Fetter Lane, E.C.4, housing Navcot, Dene Shipping, Silver Line and Seabridge.

Robert G. Crawford, former international shipping lawyer and member of S.I.H. Board, appointed Chairman, Dene Shipping and Silver Line; Renato De Paolis, Managing Director of Navcot, appointed Managing Director.

Formation of U.K.-based chemical tanker pool, Interchem Shipping Ltd., by Vlasov/Dene Group with Louis Dreyfus/Buries Markes Group, for carriage of liquid cargoes in ships of up to 15,000 tons d.w.; "Silvereagle", "Silverfalcon", "Silvermerlin" and "Silverosprey" transferred to serve in new pool.

"Thorvaldsen", 51,000 tons d.w. bulk carrier, taken on long-term bareboat charter.

"Alva Star", 228,100 tons d.w. VLCC owned by Vlasov Group and managed by Silver Line, becomes first vessel in world to undertake No. 1 special survey (hitherto always carried out in drydock) afloat.

Silver Line Golden Jubilee, 24 November.

26, 27 and 28. Three men who between them contributed not far short of a century and a half of service to Silver Line: Capt. Austin Hirst, who joined as cadet in 1927 and retired as senior master, 1973; P. A. Bull, who joined the general office of Stanley and John Thompson in 1924, played key roles in formation of Silver Chemical Tankers and Seabridge Shipping, and retired in 1973 as Silver Line's Financial Director; and Percy Rockell, who joined Thompsons six months' before P. A. Bull and who retired as Chief Accountant in 1972.

29. The 228,100 tons d.w., VLCC "Alva Star", managed by Silver Line, earlier this year became the first ship to complete Special Survey No. 1 by means of Inwater Survey, enabling her to remain in service without drydocking for a further two years.

30. One of the largest bulk carriers of her type in the world when she entered service in 1967, the 105,000 tons "Chelsea Bridge" served in the Seabridge fleet before being sold in 1972. She is seen here discharging ore at Port Talbot.

31. The 51,000 tons d.w. bulk carrier "Thorvaldsen", on long-term bareboat charter to Silver Line, pictured during sea trials. She has the capability for subsequent conversion into a roll-on/roll-off car carrier for 4,000 vehicles.

32. Robert G. Crawford Chairman, Dene Shipping and Silver Line.

a message from the Chairman

From this review of our history we can sense the vision, courage, determination and flexibility which were necessary to enable the Company to achieve its Golden Jubilee.

As we have seen, our Company has experienced and survived tremendous events, including the shipping depression in the late '20's, soon after its birth, and the holocaust of the Second World War when it lost over 60 per cent of its fleet and over 200 of its sea staff.

Now that the world's fleets are once again facing adversity, we should take our lead from the past. Such adversity is not unprecedented even in our own history and, therefore, the route having been charted, with similar qualities, together, we shall overcome the present trial and continue our expansion.

We now enter our second half century linked with the Vlasov Group.

In 1925, Stanley M. Thompson, a man of vision and energy, founded Silver Line and proceeded to develop world wide cargo liner services.

In 1937, Alexander Vlasov, an exile from his native Russia, with equal vision and energy founded his shipping group and went on to develop a fleet under several flags, including the British flag.

Today the two groups are bound together and operate a combined fleet of 47 vessels, of about $2\frac{1}{2}$ million d.w., the majority of both units and d.w. being registered under the British flag, comprising bulk carriers, tankers, combined carriers, cargo liners, chemical tankers, car carriers and passenger vessels, and operating joint ventures in several countries from East to West.

With such a past we cannot fail to succeed as did our forebears who had so much less.

We can therefore all be proud to be part of Silver Line this year and to share the responsibility of maintaining the traditions which have seen it through its first 50 years.

I ask all of you, both sea and shore staff, to join with me in adding a further page to our Company's glorious history.

33. Silver Line's new headquarters, 43 Fetter Lane, London EC4, to which it transferred last year, in company with Navcot, Dene Shipping and Seabridge.

SILVER LINE

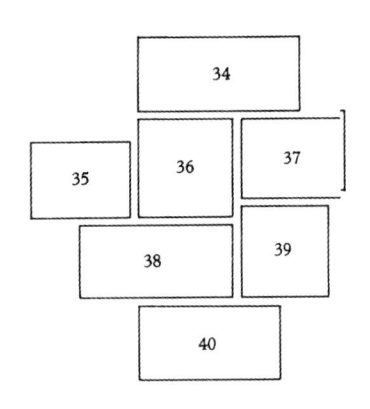

The seven Silver Line ships already in service with Seabridge, in addition to the two OBOs currently on order, represent a combined investment in the Seabridge concept of just short of 1,000,000 tons d.w.

Together with Bibby Line, another founder member, Silver Line is one of the two major participants in the consortium, the other partners in which are C. T. Bowring, Houlder Bros., and Huntings.

The overall Seabridge fleet is one of the biggest in the world, currently totalling more than 3,000,000 tons d.w. This ranges from bulk carriers of between 42,000 and 126,000 tons d.w. to OBOs of between 113,000 and 166,000 tons d.w. Almost all of the units in the fleet, which has an average age of 3.9 years, have been built specifically for Seabridge service.

The amount of cargo carried in Seabridge hulls last year exceeded 20,000,000 tons.

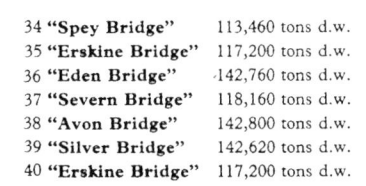

34	"Spey Bridge"	113,460 tons d.w.
35	"Erskine Bridge"	117,200 tons d.w.
36	"Eden Bridge"	142,760 tons d.w.
37	"Severn Bridge"	118,160 tons d.w.
38	"Avon Bridge"	142,800 tons d.w.
39	"Silver Bridge"	142,620 tons d.w.
40	"Erskine Bridge"	117,200 tons d.w.

Name	Delivered	Builders	Dwt.	Disposal
Silverfir	1924	Doxfords	8,255	Sunk 1941
Silverelm	1924	Doxfords	8,255	Sold 1946
Silvercedar	1924	Doxfords	8,255	Sunk 1941
Silverpine	1924	Swan Hunter	8,600	Sunk 1940
Silverlarch	1924	Swan Hunter	8,600	Sold 1947
Silveray	1925	J. L. Thompson	8,142	Sunk 1942
Silverash	1926	J. L. Thompson	8,900	Sold 1955
Silverbeech	1926	Laings	8,900	Sunk 1943
Silverbelle	1927	J. L. Thompson	8,900	Sunk 1941
Silvermaple	1927	Laings	8,900	Sunk 1944
Silverhazel	1927	J. L. Thompson	8,900	Total loss 1935
Silverguava	1927	Laings	8,927	Sold 1952
Silverpalm	1929	J. L. Thompson	9,766	Sunk 1941
Silveryew	1930	J. L. Thompson	9,766	Sunk 1941
Silverwillow	1930	J. L. Thompson	9,766	Sunk 1942
Silvercypress	1930	Harland & Wolff	10,086	Constructive total loss 1937
Silverwalnut	1930	Harland & Wolff	10,086	Sold 1954
Silverteak	1930	Harland & Wolff	10,086	Sold 1954
Silversandal	1930	Harland & Wolff	10,086	Sold 1954
Silverlaurel	1939	J. L. Thompson	8,750	Sunk 1944
Silveroak	1944	J. L. Thompson	10,376	Sold 1955
Manx Marine	Bought 1946/47	Canadian "Victory"	10,310	Sold 1948
Manx Sailor	Bought 1946/47	Canadian "Victory"	10,310	Sold 1948
Manx Navigator	Bought 1946/47	Canadian "Victory"	10,310	Sold 1949
Manx Fisher	Bought 1946/47	Canadian "Victory"	10,310	Sold 1949
Silvercedar	Bought 1947	Ex "Samsacola"	10,695	Sold 1949
Silvermaple	Bought 1947	Ex "Colorado Springs"	10,605	Sold 1952
Silverbriar	1948	J. L. Thompson	10,700	Sold 1951
Silverplane	1948	J. L. Thompson	10,700	Sold 1951
Silverholly	1949	J. L. Thompson	10,440	Sold 1949
Silveryew	1949	J. L. Thompson	10,400	Sold Ex Stocks
Silverelm	1950	J. L. Thompson	10,400	Sold Ex Stocks
Silverlaurel	1950	J. L. Thompson	10,400	Sold Ex Stocks
Silvertarn	Bought 1951	Ex "Riodene"	8,900	Sold 1954
Silverdale	1952	Lithgows	16,700	Sold 1961
Silverburn	1953	Wm. Gray	8,865	Sold 1957
Silverbrook	1953	Smiths Dock	16,630	Sold 1962
Silverdene	1956	J. L. Thompson (CR)	10,054	Sold 1960
Silverpoint	1957	Bartrams	11,700	Sold 1965
Silverfell	1957	J. L. Thompson (CR)	11,710	Sold 1963
Silverforce	1957	Laings	11,675	Sold 1964
Silverlake	1957	Laings	11,675	Sold 1963
Hallindene	Bought 1958		8,890	Sold 1958
Silversand	1958	Laings	15,465	Sold 1974
Silvercrag	1958	Laings	15,465	Sold 1972
Silverisle (Bermuda)	1960	J. L. Thompson (CR)	11,660	Sold 1965
Bishopsgate	1960	Laings	18,220	Sold 1972
Aldersgate	1960	Laings	18,220	Re-named "Silvershore" Sold 1975
Silverbeck	1961	Bartrams	13,850	Sold 1965
Silverweir	1961	J. L. Thompson	15,540	Sold 1969
Silverleaf	1963	J. L. Thompson	14,820	Sold 1968

SILVER LINE

Name	Delivered	Builders	Dwt.	Disposal
Sliverbeach (Bermuda)	1964	Ex "Totem Star" formerly "Norse Coral"	15,260	Sold 1973
Silversea (Bermuda)	1964	Ex "Totem Queen" formerly "Norse Reef"	15,260	Sold 1973
Tower Bridge	1965	Laings Ex "Silver How"	34,405	Sold 1970
Silverkestrel	1965	Ekeroths Norrkoping	648	Sold 1975
Silverfalcon	1966	Lodose Varf	1,900	
Silvercove	1967	Namura Shipyard	18,656	
Silvercape	1967	Namura Shipyard	18,613	Sold 1972
Sigsilver (Chelsea Bridge)	1967	Ishikawajima Harima	105,779	Sold 1973
Gallic Bridge	1967	Lithgows	75,350	Sold 1974
Silvermerlin	1968	Lodose Varf	1,871	
Bellnes (Silverforth)	1969	Lithgows	19,710	
Silvereid	1969	Hall Russell	2,470	Sold 1974
Silverhawk	1969	Cammell Laird	10,392	
Spey Bridge	1969	Sumitomo	113,460	
Silvermain	1970	Brodogradiliste	25,541	
Silverosprey	1970	Cammell Laird	6,260	
Silverharrier	1970	Hall Russell	6,162	
Binsnes (Silvertweed)	1970	Lithgows	21,206	
Baknes (Silverclyde)	1970	Lithgows	21,206	
Silvereagle	1970	Cammell Laird	6,260	
Avon Bridge	1971	Sumitomo	142,801	
Silverfjord	1972	Brodogradiliste	28,420	
Eden Bridge	1972	Sumitomo	142,762	
Stirling Bridge	1972	Mitsubishi	118,288	
Severn Bridge	1972	Mitsubishi	118,167	
Silver Bridge	1972	Sumitomo	142,627	
Erskine Bridge	1973	Mitsubishi	117,200	
Silverdon	1973	Lithgows	32,300	
Silverpelerin	1973	Krogerwerft	6,850	

SILVER LINE MANAGED TONNAGE

Name	Delivered	Builders	Dwt.	Disposal
Altanin	1964	Deutsche Werft	88,513	
Alva Star	1969	Gotaverken	228,100	
Alva Bay	1973	Gotaverken	222,331	
Alva Sea	1973	Gotaverken	221,457	
Thorvaldsen	1974	Burmeister & Wain	51,999	

NEWBUILDINGS

Name	Delivered	Builders	Dwt.	Disposal
Product Carrier	1976/7	Cammell Laird	55,000	
Product Carrier	1976/7	Cammell Laird	55,000	
Product Carrier	1976/7	Cammell Laird	55,000	
Product Carrier	1976/7	Cammell Laird	55,000	
Product Carrier	1976/7	Cammell Laird	55,000	
Ore/Bulk/Oil Carrier	1977	Mitsubishi	120,000	
Ore/Bulk/Oil Carrier	1977	Mitsubishi	120,000	

FERRIES OF THE AMAR LINE

In 1975, to extend the Group's activities and also to face the oil crisis, Vlasov decided to open in Jeddah the first Saudi Arabian-headquartered Western shipping concern, the Amar Line. This company, which ceased activities ten years later, was initially a fifty/fifty joint venture between the Vlasov Group and the Arabian Maritime Co. owned by Gaith Pharaon, a Saudi Arabian entrepreneur who already had several business relationships with American and European companies and was well-known in the industrial Western World.

Apart from transferring to Amar Line three freighters and three tankers owned by Vlasov and buying a landing craft to assist them in Jeddah (see *Marland II*, page 43), it was decided to open a regular car and passenger ferry route between Jeddah and Suez. For this purpose in 1976 the Amar Line's subsidiary Saudi Maritime Transport Co. started service with the blessing of the Arab League. The first vessel owned by the new concern was the *Saudi Moon*.

She left Jeddah on her first voyage to Suez on 15th June 1976. On the outward voyage she was filled to capacity with second-hand cars. At the time, in fact, modern cars (particularly American and Japanese makes) were very popular in Saudi Arabia which was a great importer. Thus, the second-hand car market was also booming, with Egypt and its neighbouring countries being the main buyers. On the return voyage the *Saudi Moon* and her running mates transported North African pilgrims to Mecca and, as many of them were travelling with their own camel or (in the case of the richest) horse to continue the journey via land from

Side elevation of the *Saudi Moon*, the vessel with which Amar Line started its ferry operation between Jeddah and Suez in 1976.

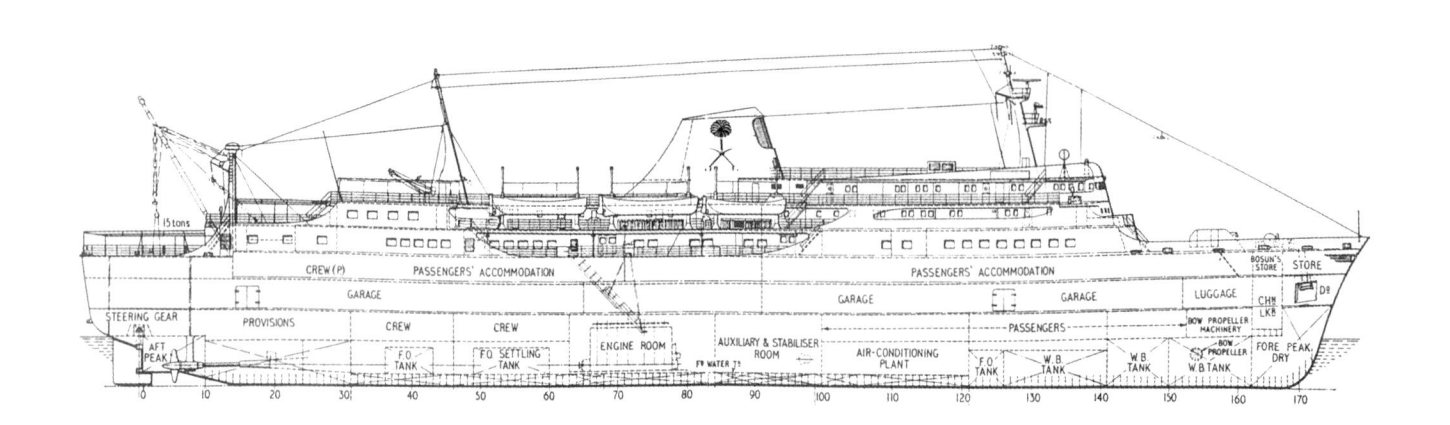

Jeddah, the ro/ro door and the garage were perfectly suitable for their needs and the latter was quickly transformed into a comfortable stable for the animals. A colourful description of these voyages is to be found in Capt. Guido Badano's book "Ricordi di un Capitano"; he was master on Saudi Maritime Transport Co.'s ferries between 1980 and 1984.

In 1977 Gaith Pharaon sold his 50% holding in the Amar Line; 20% were taken over by Vlasov, who thus took control, while the remaining 30% were purchased by Al Rajhi-Al Jedais Company-Riyadh.

In 1979 the line was so successful that Kloster's former first cruise ship, *Sunward*, was bought and re-named *Saudi Moon I* to partner the *Saudi Moon*. But during the same year the Arab League forced the latter to be withdrawn from service: a group of Muslin pilgrims, upset at finding everywhere on board signs in Hebrew, had formally protested at the use of a former Israeli vessel to bring the pilgrims to Mecca.

As a replacement it was decided to buy the modern Greek ferry *Oinoussai*, re-named *Belkis I* (Arabian Princess). From March 1980, the *Saudi Moon I* and *Belkis I* started to call also at Aquaba and Port Sudan.

The Saudi Maritime Transport Co. ferries, crewed and managed by Vlasov's Shipping Management S.A.M. (later V.Ships), had their time schedule co-ordinated with the Arabian Ferry Saudi Maritime Agencies of Jeddah and were partnered on the same route by the *Sindibad* and *Sharazad*.

Side elevation of the *Saudi Moon I*; originally delivered as Kloster's *Sunward*, her dimensions and high standard made of her the flagship of the Amar Line ferry service in the Red Sea.

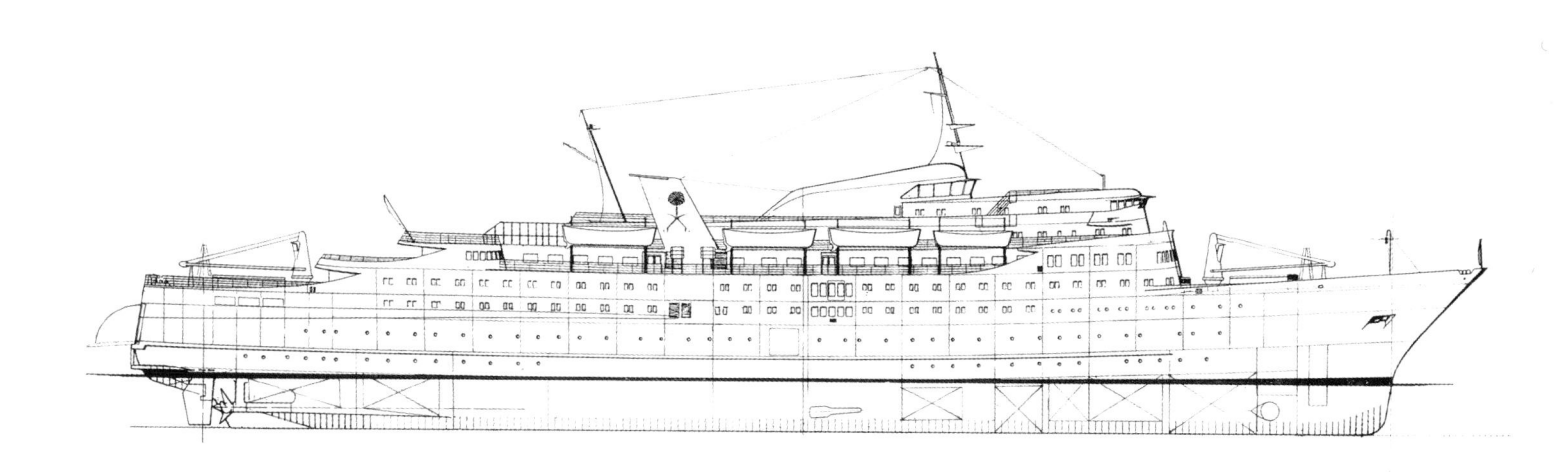

FERRIES OF THE AMAR LINE

SAUDI MOON (1976-1979)

Ro/ro passenger ferry
(formerly *Bilu*, *Dan*, *El Greco*, then *Golden Sky*, *Vergina*, *Mir*)
Builders: Cockerill-Ougrée (Hoboken, Belgium)
8039 grt (3447 nrt) 1612 dwt
127.79 [117.20] x 18.51 x 5.12 m
(419.3 [384.5] x 60.7 x 16.8 ft)
2 single-acting 7-cyl. diesels; 10500 BHP; 18 kn
by Fiat Grandi Motori (Turin)
524 berths; 350 deck passengers

1964: launched as *Bilu* for Bilu-Somerfin Car Ferries Ltd, Haifa, Israeli flag.
1967: sold to Kavim Hevrat-Oniyot B.M., and re-named *Dan*.
1976: sold to the Greek shipowner Potamianos and re-named *El Greco*; p.o.r. Piraeus; remained laid up. Re-sold after a few months to Amar Line's subsidiary Saudi Maritime Transport Co., Jeddah and renamed *Saudi Moon*; Saudi Arabian flag.
Entered service on 15th June on the Jeddah-Suez ferry route.
1979 May: laid up and re-named *Golden Sky*; put up for sale.
Sold to Vergina Ferry, Piraeus, a subsidiary of Stability Maritime Corp. (Axios Efthimios) and re-named *Vergina*.
1998: bare-boat chartered with a purchase option to a Ukrainan company for ferry service in the Black Sea. August: after machinery troubles, returned to her owners and laid up.

SAUDI MOON I (1979-1985)

Ro/ro passenger ferry
(formerly *Sunward*, *Île De Beauté*, *Grand Flotel*, then *Ocean Spirit*, *Scandinavian Song*, *Santiago de Cuba*, *The Empress*)
Builders: Mekanisker Verksted (Bergen)
10558 grt (3741 nrt) 1724 dwt
139.42 [122.50] x 20.80 x 5.26 m
(457.4 [401.9] x 68.2 x 17.2 ft)
2 B&W 2 stroke single-acting 12-cyl. diesels
13200 BHP; 20 kn
by B&W (Copenhagen)
350 berths; 634 deck passengers

1965 September 11th: keel laid.
1966 March 24th: launched as *Sunward* for Klosters Sunward Ferries Ltd, a subsidiary of Klosters Rederi A/S, Oslo. On the following 20th June she was delivered, entering service five days later as a passenger/car ferry on the Southampton-Vigo-Lisbon-Gibraltar line. In November the *Sunward* was withdrawn from service owing to poor bookings and transferred to Kloster's newly established Norwegian Caribbean Lines. Re-entered service on 19th December from Miami for 3- and 4-day cruises to Nassau.
1973 January: sold to the French Cie. Générale Transméditerranéenne and sent to a Marseilles shipyard to be adapted for the Toulon-Porto Torres (Sardinia) ferry service. Re-entered service in July with the new name of *Île De Beauté*.
1976 January: transferred to the new Société Nationale Corse-Méditerranée (into which her for-

*In 1980 Amar Line bought the Greek ferry Oinoussai and re-named her Belkis I; they put her on the Jeddah-Suez ferry line as a replacement for the Saudi Moon and as a running mate to the Saudi Moon I.
(Alberto Bisagno, Genoa)*

Ferries of the Amar Line

mer owners were merged) and employed as a French exhibition ship in the Middle East. Later that year sold to Eastern Gulf Inc., Panama, renamed *Grand Flotel* and permanently berthed at Sharajah as an accommodation ship.
1979 April: sold to Amar Line's subsidiary Saudi Maritime Transport Co. of Jeddah, which put her in service as a running-mate to the *Saudi Moon*, with the name of *Saudi Moon I*.
1985: chartered to Sabah Maritime Services Co., Jeddah which maintained her on the same route with the same name; management entrusted to Saudi Maritime Agencies of Jeddah.
1988: chartered to Ocean Quest International, New Orleans, registered at Nassau as *Ocean Spirit* and sent to the Sembawang shipyard of Singapore to be converted into a cruise ship for diving enthusiasts.
1989 March 4th: re-entered service from New Orleans for cruises to Cozumel, Belize and Cancun.
1990 November: sold to Ferry Charter Florida Inc.
1991 January: chartered to SeaEscape of Miami and re-named *Scandinavian Song*. After a brief use for cruises from St. Petersburg she was permanently based in Miami. In November the *Scandinavian Song* was trasferred to Danish Cruise Line, a joint-venture between SeaEscape and Nordisk Faergefart, for daily cruises from San Juan to St. Thomas.
1993: back to SeaEscape and employed for daily cruises from Port Canaveral.
1994: chartered to F.lli Cosulich, Genoa, re-named *Santiago de Cuba* and briefly used for cruises from Cuban ports. This attempt, however, did not meet the hoped-for response and later that year she was dispatched under charter to Penang with the new name of *The Empress* for cruises in Far East waters. On 22nd November collided with the tanker *Ocean Success* and put back to Singapore to be repaired.
1998: in service.

BELKIS I (1979 -1985)
Ro/ro passenger ferry
(formerly *Oinoussai*,
then *Al Shahba*, *Mena*, *Al Loloa*)
Builders: Argo S. B. & Repairing Co. (Perama)
3114 grt (1775 nrt) 1016 dwt
92.56 [81.23] x 14.06 x 4.10 m
(303.7 [266.5] x 46.1 x 13.4 ft)
2 8-cyl. Diesel engines; 17200 SHP; 17,5 kn
by Stork-Werkspoor (Amsterdam)
180 berths; 860 deck passengers

1972 April 1st: launched with the name *Oinoussai* for Oinoussai Shipping Corp. (L. Karras & K. Pontikos), Piraeus.
1973 July: completed and entered service.
1979 November: bought by Vlasov's Acomarit UK Ltd, bare-boat chartered to Amar Line's subsidiary Saudi Maritime Transport Co. of Jeddah as a replacement for *Saudi Moon* to work along with the *Saudi Moon I* in the Red Sea; Vlasov's Shipping Management S.A.M. appointed managers. Refitted; original 2593 grt increased to 3114.
1984 July 5th: ran aground while en route from Suez to Haruba. Freed with the assistance of a tug and brought to Jeddah where she was laid up.
1987 July 8th: Lloyd's Register class withdrawn.
1990: sold to International Trading Co., Jeddah and re-named *Al Shahba*.
1992: laid up, name changed to *Mena* and put up for sale. Bought by Raafor-High Seas of Panama and re-named *Al Loloa*; again employed on the Jeddah-Suez ferry route.
1994 July 12th: fire in the engine room while en route from Suez to Jeddah, which later engulfed the whole vessel. Sank the following day six miles north of Safaga, in position 27°02'N-03°43'E.

The *Saudi Moon I* was the largest and most modern ferry employed by Amar Line; the company used as funnel logo two crossed scimitars from which rose a palm tree in golden yellow on a green background.
(Alberto Bisagno, Genoa)

V.Ships Today & Tomorrow

by Per Bjornsen, V.Ships Leisure Project Development Manager

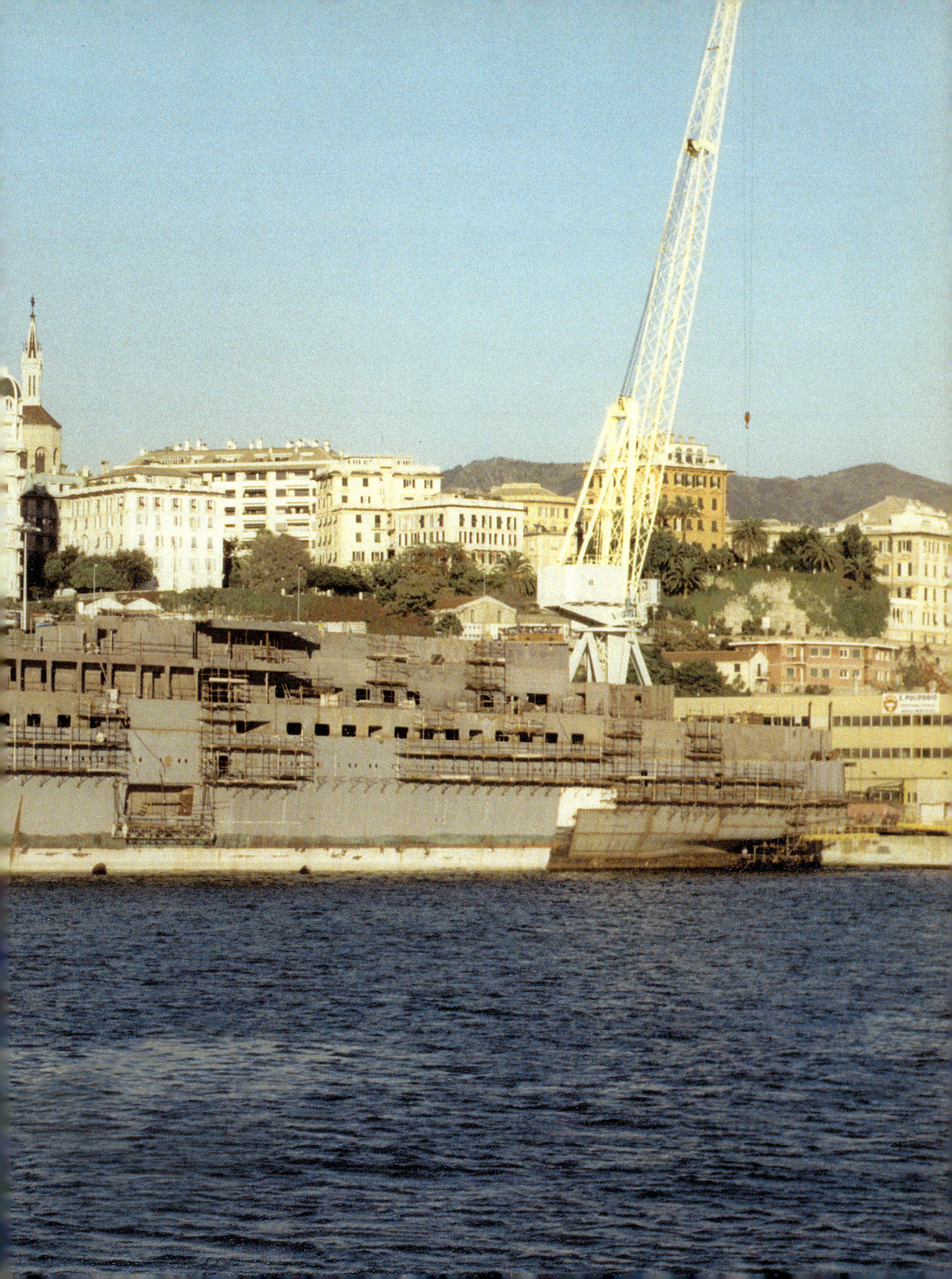

V.Ships Today & Tomorrow

Previous pages: the latest newbuilding for the V.Ships' fleet, the *Seven Seas Navigator* under construction at Genoa. *(Enrico Repetto, Genoa)*

Below: the "Aigue Marine" building in Monaco Monte-Carlo, where the Vlasov Group headquarters are presently located.

As we approach the millennium, V.Ships has become the World's largest independent ship management company with a fleet of approximately 400 vessels.
The company is also the leading provider of third party ship management services to the cruise industry.
The company provides services to more than 80 clients world-wide through 25 offices, has a roster of more than 10,000 seagoing staff and a shore based staff of about 550.
The Group's activities are divided into four divisions:
- the **Ship Management Division** which is the largest one. The technical and crew management of cargo vessels continues to be the Group's core business activity;
- the **Commercial Division** which comprises V.Ships' specialist chartering and sale & purchase company (Silver Line Ltd), as well as ship agency activities (Seamaster);
- the **Leisure Division** which provides ship management, chartering, project management and media services to the cruise industry;
- the **Corporate Services Division** which consists of insurance, financial brokerage, marine legal services and average adjusting.

The beginning

The company started in 1962 under the name of Shipping Management to provide ship management services to the Vlasov fleet. In the 'seventies, Shipping Management started to provide services to a selected group of business partners, amongst others Docenave in Brazil and Tradax in Switzerland. However, the Group also wanted to diversify its shipping activities and, as there was a large demand from other shipowners for the same service, it started to offer this to other clients.
The Group expanded its management activities under the name of International Shipping Management, offering its services to financial and ship owning groups. In 1984 the growth in the number of ships under management for third parties led to the formation of an independent service division with the trade name of V.Ships.

V.Ships Today & Tomorrow

Shareholders

V.Ships, originally a part of the Vlasov Group of companies, now has three separate shareholders. In 1986 it already had 80 vessels under management. Boris Vlasov wanted to reward top management and give them an incentive to continue its development. He did this by selling them fifty per cent of the shares, the remainder being held by the Vlasov Group.

In 1993 GE Capital, the financial arm of General Electric, wanted to invest in a ship management company to diversify its shipping investment portfolio. By nominating its "in-house" ship manager it would have better control of its own shipping finance projects. It contacted V.Ships about the possibility of acquiring a minority share. The Vlasov Group and the directors agreed to sell ten percent each, so that today the Vlasov Group retains a 40% stake, senior management a further 40% whilst last 20% is held by GE Capital.

The company has had a stable management and loyal staff. The Group's

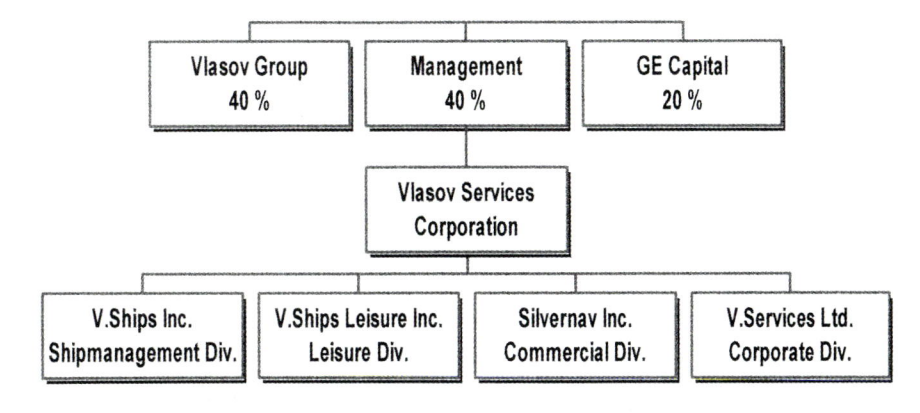

Above: V.Ships' stand during the 1998 Mediterranean Seatrade convention held in Genoa.

Below: The V.Ships shipping management and technical department offices and (left) the shareholders diagram.

V.Ships Today & Tomorrow

Chairman and President are today as in 1984 Mauro Terrevazzi and Tullio Biggi respectively. A large proportion of the staff has been working for the Group since the early years.

The Ship Management Division
(headed by Herman Messner)
The Vlasov Group had been represented in New York since 1947 through their company Navcot, but in 1984 the office was renamed V.Ships (USA) Inc. In 1996 the office merged with International Marine Carriers (IMC), moved from Manhattan to IMC's offices in Mineola, Long Island and was re-named V.Ships Marine Inc.

The company's second branch office was V.Ships (U.K.) Ltd in Southampton which opened in 1989, followed in the same year by V.Ships (Norway) and V.Ships (Dubai).

In 1990 V.Ships opened an office in Cyprus, followed by that of Rio de Janeiro in 1994 which was a rename of the Group's Atlas Partecipaçaos office. The Asia Pacific office was opened in Singapore in 1996 whilst a second office, V.Ships (Singapore) for management of cargo vessels started activity in 1997. V.Ships (Asia Pacific) offers cruise related services. The latest office opened is V.Ships (Florida) Inc. in Miami.

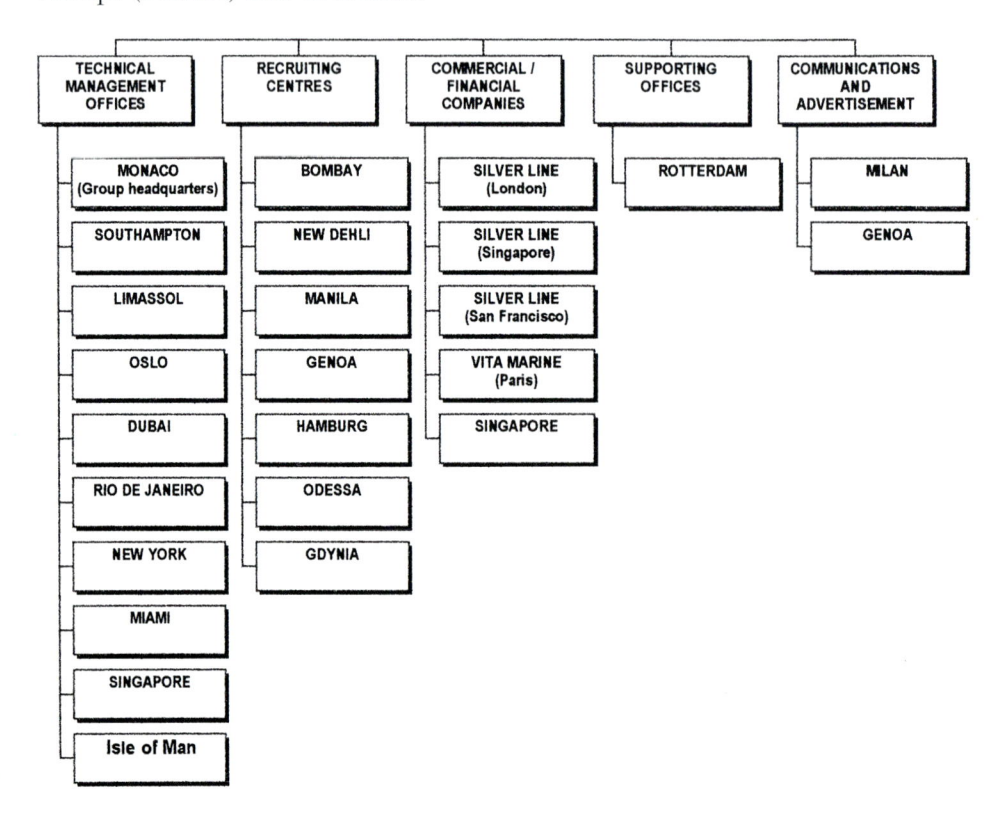

The V.Ships leisure department office.

A diagram showing the today's V.Ships organisation

TECHNICAL MANAGEMENT OFFICES	RECRUITING CENTRES	COMMERCIAL / FINANCIAL COMPANIES	SUPPORTING OFFICES	COMMUNICATIONS AND ADVERTISEMENT
MONACO (Group headquarters)	BOMBAY	SILVER LINE (London)	ROTTERDAM	MILAN
SOUTHAMPTON	NEW DEHLI	SILVER LINE (Singapore)		GENOA
LIMASSOL	MANILA	SILVER LINE (San Francisco)		
OSLO	GENOA	VITA MARINE (Paris)		
DUBAI	HAMBURG	SINGAPORE		
RIO DE JANEIRO	ODESSA			
NEW YORK	GDYNIA			
MIAMI				
SINGAPORE				
Isle of Man				

V.Ships Today & Tomorrow

In addition, the Group has recruiting centres in Bombay, New Delhi, Manila, Genoa, Hamburg, Odessa and Gdynia. The crew roster of deck & engine personnel consists mainly of Indian, Filipino, Ukrainian and Polish nationals.

The Group manages various types of vessel: passenger vessels, gas carriers, ro-ro, cable layers, tween deckers, chemical carriers, container vessels, OBOs, bulk carriers, product carriers and ferries.

In 1998 the company merged its ship management and shipping service activities with Celtic Marine to create one of the World's largest maritime service groups. The manning and management services of Celtic Marine have been incorporated in V.Ships. The new concern is the World's largest ship management company with a fleet of approximately 400 vessels.

Silver Wind (foreground) and *Silver Cloud* being fitted out in Genoa; they are one of the best recent examples of "full concept management" offered by V.Ships. (Paolo Piccione, Genoa)

The Commercial Division
(headed by Tony Crawford)

Silver Line was founded in 1925, but became a part of the Vlasov Group in 1973 with the acquisition of its parent company, Shipping Industrial Holdings Ltd, the UK's second-largest shipping business.

The Vlasov-owned *Albatros* and the *Radisson Diamond* passing each other in the Panama Canal; both cruise vessels are managed by V.Ships.

V.Ships Today & Tomorrow

Up to 1984 the company provided commercial ship management services to the Vlasov Group's vessels, but since its conversion into V.Ships it provides these services to third party clients, whether the ships are managed by V.Ships or not. Silver Line is today the commercial division of the company and has some 100 vessels under commercial (sale & purchase and chartering) management. The head office is in London with branch offices in Singapore and San Francisco, opened in 1997 and 1998, respectively.

In 1993 Seamaster Shipping and Trading was founded as a fully owned subsidiary to provide ship agency services to the V.Ships fleet. The company expanded in 1995 by opening of Seamaster France, in 1996 with that of Seamaster Singapore and in 1997 the office in Belgium. Seamaster acts as agents for Pegasus Ship Repair Services Inc., which provides riding squad repair teams.

The Leisure Division
(headed by Roberto Giorgi)

V.Ships established a separate Leisure Division in 1993 to re-enter the cruise vessel management sector, taking advantage of its experience as previous owners of Sitmar Cruises. It is the only management company that offers a "full concept" ship management service to the cruise industry. In addition to traditional deck and engine management, it provides hotel, catering and entertainment management as well as assistance with revenue earning activities.

In 1996, V.Ships Leisure S.A.M. was established as a separate company based in Monaco. The company supplies passenger ship management with services

Ken Norman and Arnold Brereton, senior naval architects and marine engineers of the V.Ships newbuilding department and main designers of the Seven Seas Navigator.

A drawing showing the intended final aspect of the Seven Seas Navigator, when completed in Summer 1999.

from Monaco, Mineola (New York) and Singapore. V.Ships Leisure has supervised and managed a number of prestigious cruise projects, including the creation of Silversea Cruises as well as the negotiation and supervision of the newbuilding contract for their vessels *Silver Cloud* and *Silver Wind*.

V.Ships provided Italian officers to Fearnley & Eger, the original owner of Renaissance Cruises, and provided full management services to the financing banks until the company was sold to the Cameli Group.

The company also provided full management services to the fleet of the then new British cruise line Airtours Sun Cruises during the important first years of its operation between 1995 and 1998.

One of the vessels presently under full management is the *Minerva*, an ex-Soviet research vessel acquired by the Group and converted into a four-star cruise vessel. The contracts with the shipyard and the Charterers, P&O' subsidiary Swan Hellenic Cruise, were negotiated by the Newbuilding Department of V.Ships.

The Leisure Division manages the *Albatros*, which is on charter to the German operator Phoenix Reisen.

In 1998 the Vlasov Group entered into a joint venture with Radisson Seven Seas Cruises for the construction of a 500-passenger luxury vessel, the *Seven Seas Navigator*. As part of the same venture the RSSC vessels *Radisson Diamond* and *Song of Flower* have been taken in technical management by V.Ships and the company will also provide manning services for hotel crew on board their *Paul Gaugin*.

In 1997 V.Ships established itself in media, advertising and communications with the creation of Trans World Communications. The company has offices in Milan and Genoa, and has a number of clients in the cruise and leisure industry.

A mock-up for the standard suite of the *Seven Seas Navigator* and the aspect of the vessel in October 1998 while undergoing her transformation.

The Corporate Services Division

Marine Legal Services was established in 1996 as an independent legal consultant to V.Ships and its clients as well as to the shipping industry in general.

Vanguard Adjusting was also established in 1996. This is an average adjusting company dealing with all types of insurance claims, including loss of hire, cruise indemnity and collision recoveries.

Vita Marine was established in 1990 as an independent ship finance "boutique" in Paris, because there was a demand from investor clients for "package" shipping deals where the Group presented all aspects of service including acquisition of vessel, financing, securing of time charter contract and ship management.

The author (second from right) with the V.Ships newbuilding team.

V.Ships tomorrow

The company is growing steadily and the fleet has quadrupled in size over the last 10 years.

It will continue to increase its international network through new offices, joint ventures, mergers and acquisitions.

It aims to be the World's number one provider of a spectrum of services as broad as possible in order to be able to offer "Global Shipping Services".

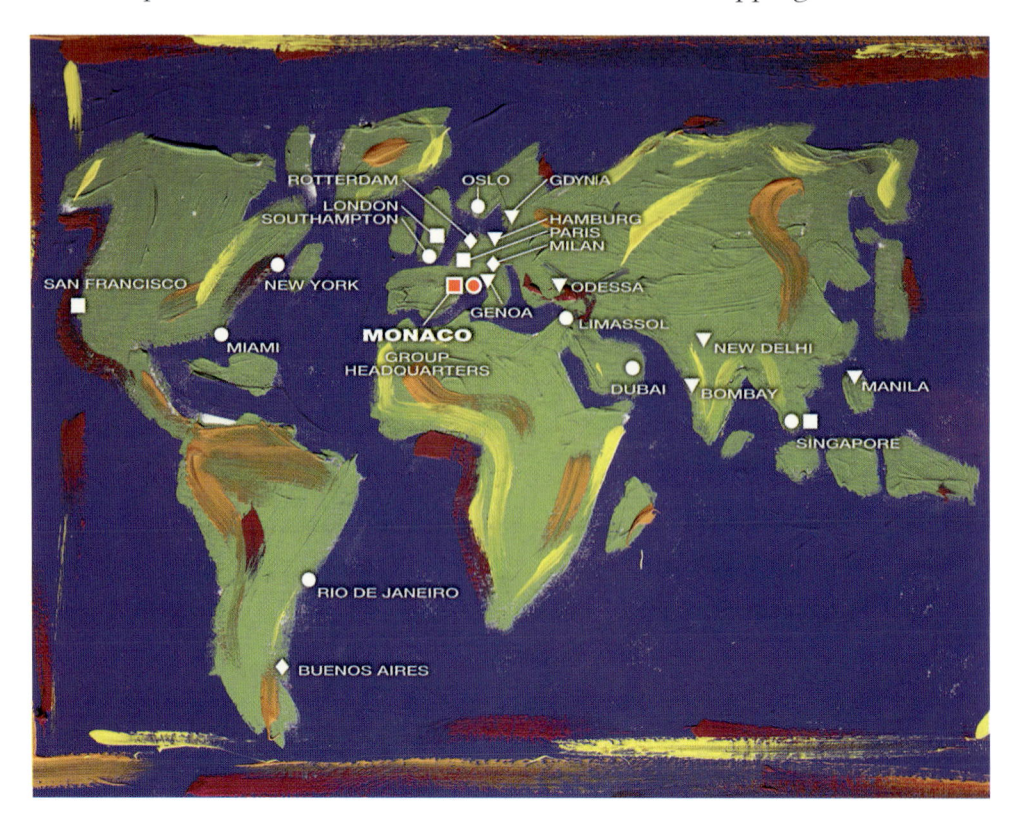

V.Ships offices in the World.

V.Ships Highlights

1962   Formation of Shipping Management, Principality of Monaco

1984   Formation of V.Ships Inc. as separate division and V.Ships as a trade name
35 ships under management

1986   Management buy-out of 50 per cent
80 ships under management

1989   Opening of V.Ships (U.K.) Ltd, Southampton
Opening of V.Ships (Norway) AS
Opening of V.Ships (Italy) Srl, Genoa
Opening of V.Ships (Dubai)

1990   100 ships under management
Opening of V.Ships (Cyprus) Ltd
Formation of Vita Marine, Paris

1993   GE Capital acquires 20 per cent of the Group
Formation of V.Ships Leisure Division
Formation of Seamaster Shipping and Trading, Holland
Renaming of Atlas Pacific to V.Ships (Rio) Inc., Rio de Janeiro

1995   Opening of Seamaster (France)

1996   200 ships under management
Opening of V.Ships (Asia Pacific) Pte. Ltd, Singapore
Opening of Seamaster Agency Services, Singapore
Merger V.Ships (USA) Inc. and International Marine Carriers to form
V.Ships (Marine) Inc., Mineola

1997   Formation of V.Ships Leisure S.A.M., Principality of Monaco
Opening of Seamaster Belgium
Formation of Transworld Communications
Opening of Silver Line (Singapore) Pte. Ltd,
Opening of V.Ships (Singapore) Pte. Ltd.
Opening of V.Ships (Florida) Inc., Miami

1998   Opening of Silver Line (USA) Ltd, San Francisco
Merger of V.Ships and Celtic Marine
400 vessels under management

Emigrant Liners Deck Plans

"SITMAR"
SOCIETÀ ITALIANA TRASPORTI MARITTIMI S.p.A.

	Spazi di ricreazione / Deck recreation Space / Cubierto de recreo
	Sale e passaggi / Public Rooms & Spaces / Salones y pasillo
	Cabine / Cabins / Camarotes
	Dormitori / Dormitories / Dormitorios
	Servizi Generali / General Services / Servicios generales

T/n "CASTEL BIANCO"

PIANO SISTEMAZIONI PASSEGGERI
PASSENGER ACCOMMODATION PLAN
PLAN SISTEMACIONES PASAJEROS

LEGGENDA	GUIDE PLAN	EXPLICACION
Letti normali	Normal berths	Camas normales
Letti longhi	Wide berths	Camas anchas
A-C-E Letti bassi	Lower berths	Camas bajas
B-D-F Letti alti	Upper berths	Camas altas
Bagno	Bath	Baño
Doccia	Shower	Ducha
W.C.	W.C.	W.C.
Lavandino	Wash Basin (W.B)	Lavatorio
Guardaroba	Wardrobe (W.R)	Ropero
Spazio per giuochi	Games deck	Cubierto de juegos
Sala Soggiorno	Lounge	Salon de Recreo
Piscina	Swimming pool	Pileta
Bar	Bar	Bar
Sala Signore	Ladies parlour	Salon señoras
Salone da pranzo	Dining saloon	Salon comedor
Passeggiata	Promenade	Paseo
Sala scrittura	Writing room	Salon de corrispond
Sala bambini	Children's room	Salon para niños
Parr. per Signore	Hairdresser	Peluqueria para Señoras
Barbiere	Barber shop	Peluquero
Uff. commissario	Purser's office	Oficina del Comisario

PONTE "SOLE" / "SUN" DECK / PUENTE DEL SOL

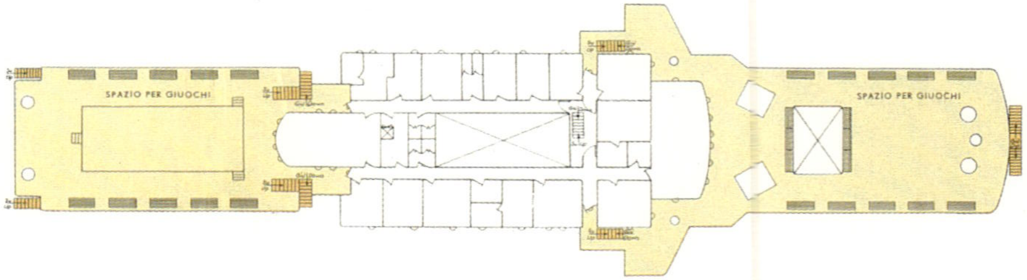

PONTE IMBARCAZIONI / BOAT DECK / PUENTE DE BOTES

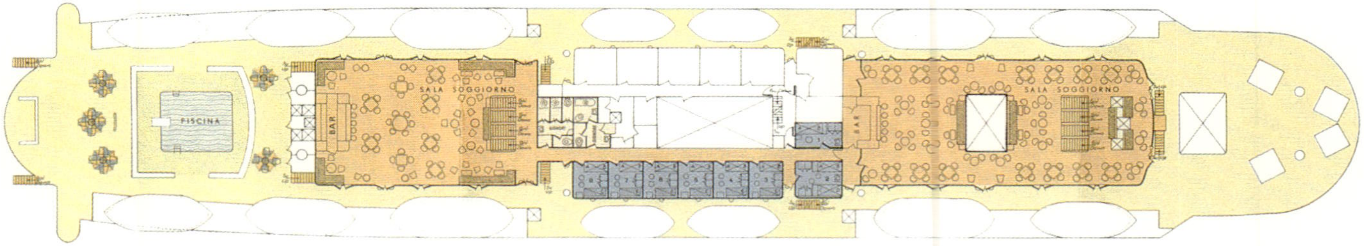

204

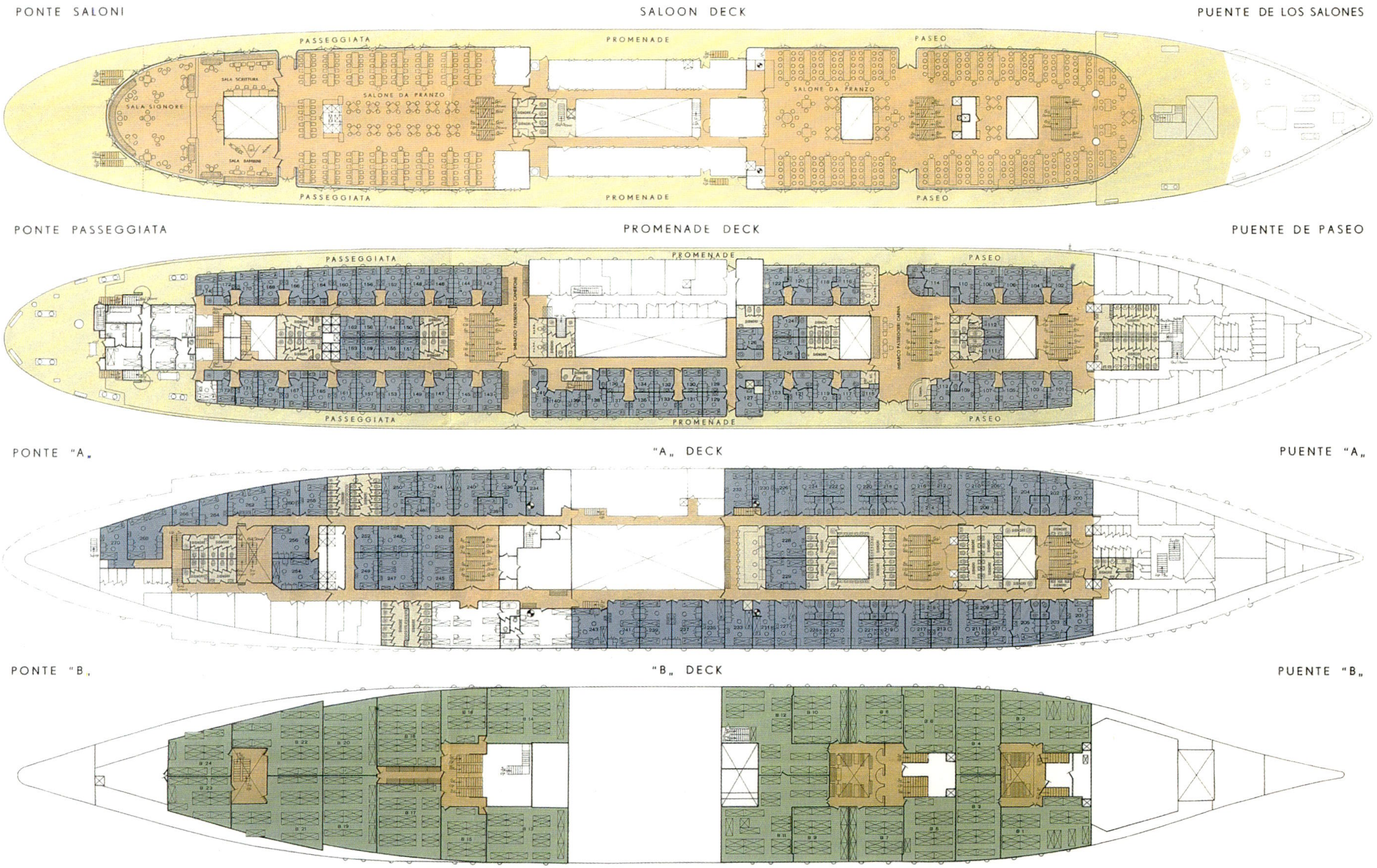

CAPACITÀ POSTI IN CABINE

PONTE IMBARCAZIONI:

6 Cabine Esterne × 3 posti = 18 posti
1 Cabina Esterna × 4 " = 4 "
1 " Interna × 2 " = 2 "
8 Cabine con . . . . **24** posti

PONTE PASSEGGIATA

27 Cabine Esterne × 2 posti = 54 posti
1 Cabina Esterna × 3 " = 3 "
28 Cabine Esterne × 4 " = 112 "
18 " Interne × 2 " = 36 "
74 Cabine con . . . . **205** posti

Delle quali

5 Cabine con bagno privato = 11 posti
4 " associabili . . . . . " = 16 "
16 " con WC e doccia " = 32 "
11 " associabili . . . . . " = 38 "
36 Cabine . . . . **97** posti

PONTE "A":

3 Cabine Esterne × 2 posti = 6 posti
4 " " × 3 " = 12 "
29 " " × 4 " = 116 "
9 " " × 6 " = 54 "
7 " Interne × 2 " = 14 "
4 " " × 4 " = 16 "
5 " " × 6 " = 30 "
61 Cabine con . . . . **248** posti

Delle quali

14 cabine con WC e doccia privata = 50 posti
6 " associabili . . . . . . " = 24 "
20 Cabine . . . . = **74** posti

RIEPILOGO POSTI IN CABINA:

Ponte Imbarcazioni 8 cabine = 24 posti
" passeggiata 74 " = 205 "
" "A" 61 " = 248 "
Totale 143 cabine con 477 posti

Delle quali

5 Cabine con privato per posti 11
4 " associabili . . . . " " 16
30 " con privato " " 82
17 " associabili . . . . " " 62
35 cabine con locale d'igiene privato = " 93
21 " associabili . . . . = " 78
Totale posti con locali d'igiene privato o associato = " 171

Posti in Cabine = 477
" " Dormitori = 717
TOTALE COMPLESSIVO **1194** POSTI

Berths in Cabins = 477
" " Dormitories = 717
GRAND TOTAL BERTHS **1194**

CAPACITÀ POSTI IN DORMITORI

Dormitorio	**B** 1	= 36	posti di cui	—
"	**B** 2	= 36	" " " "	
"	**B** 3	= 32	" " " "	4
"	**B** 4	= 32	" " " "	2
"	**B** 5	= 24	" " " "	4
"	**B** 6	= 24	" " " "	4
"	**B** 7	= 26	" " " "	2
"	**B** 8	= 26	" " " "	2
"	**B** 9	= 26	" " " "	2
"	**B** 10	= 26	" " " "	2
"	**B** 11	= 38	" " " "	38
"	**B** 12	= 34	" " " "	34
		360	posti di cui	**94**

Camerone	**B** 13	= 35	posti di cui	—
"	**B** 14	= 34	" " " "	—
"	**B** 15	= 18	" " " "	—
"	**B** 16	= 18	" " " "	—
"	**B** 17	= 42	" " " "	—
"	**B** 18	= 42	" " " "	—
"	**B** 19	= 30	" " " "	8
"	**B** 20	= 30	" " " "	8
"	**B** 21	= 34	" " " "	4
"	**B** 22	= 32	" " " "	9
"	**B** 23	= 22	" " " "	3
"	**B** 24	= 20	" " " "	3
Totale		**357**	posti di cui	35

Totale posti in dormitorio 717 di cui 129

Puestos en camarotes = 477
" " dormitorios = 717
TOTAL GENERAL PUESTOS **1194**

STAMPATO IN ITALIA - PRINTED IN ITALY
ARTI GRAFICHE G. SCHENONE - GENOVA

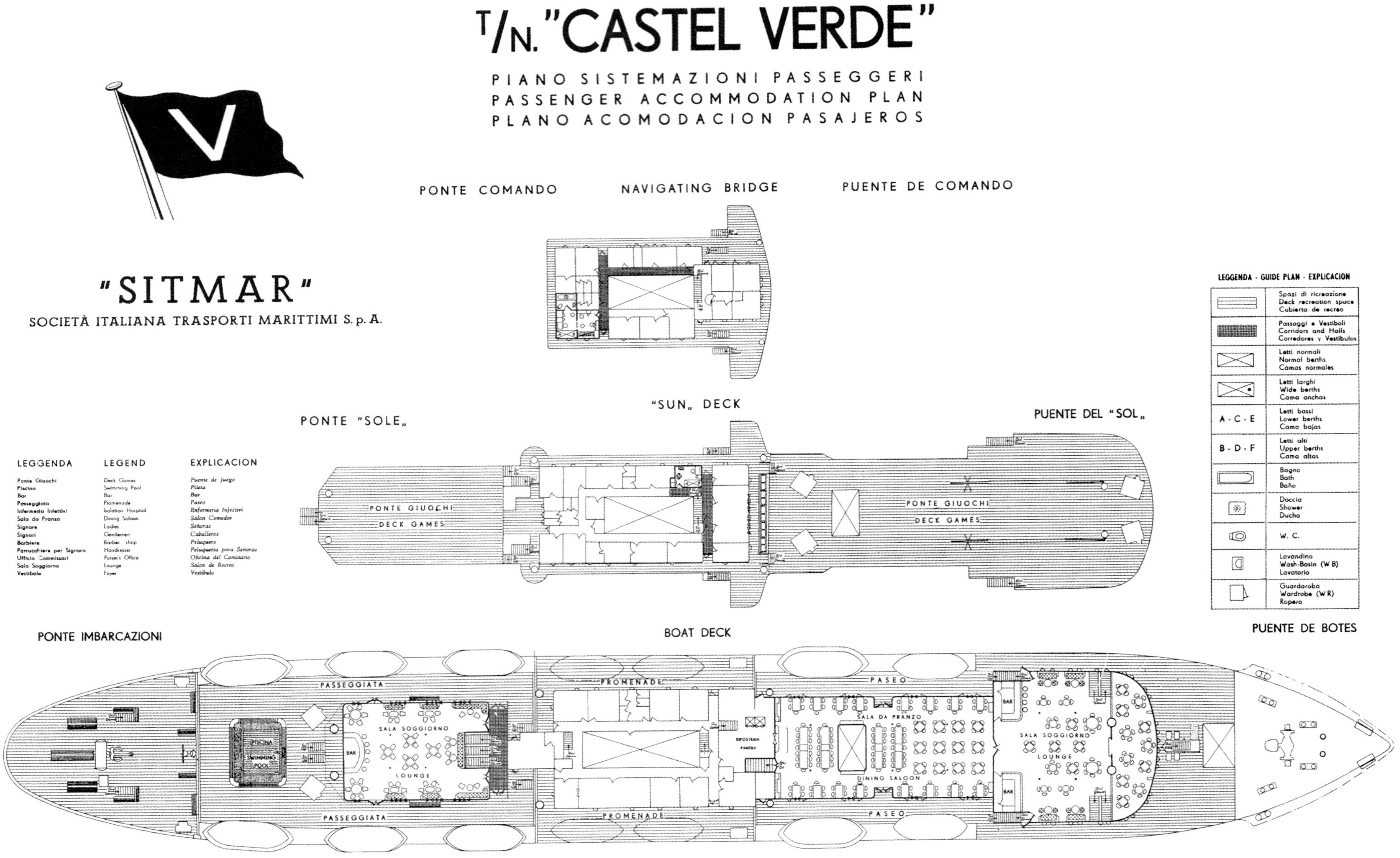

T/N. "CASTEL VERDE"

PIANO SISTEMAZIONI PASSEGGERI
PASSENGER ACCOMMODATION PLAN
PLANO ACOMODACION PASAJEROS

"SITMAR"
SOCIETÀ ITALIANA TRASPORTI MARITTIMI S.p.A.

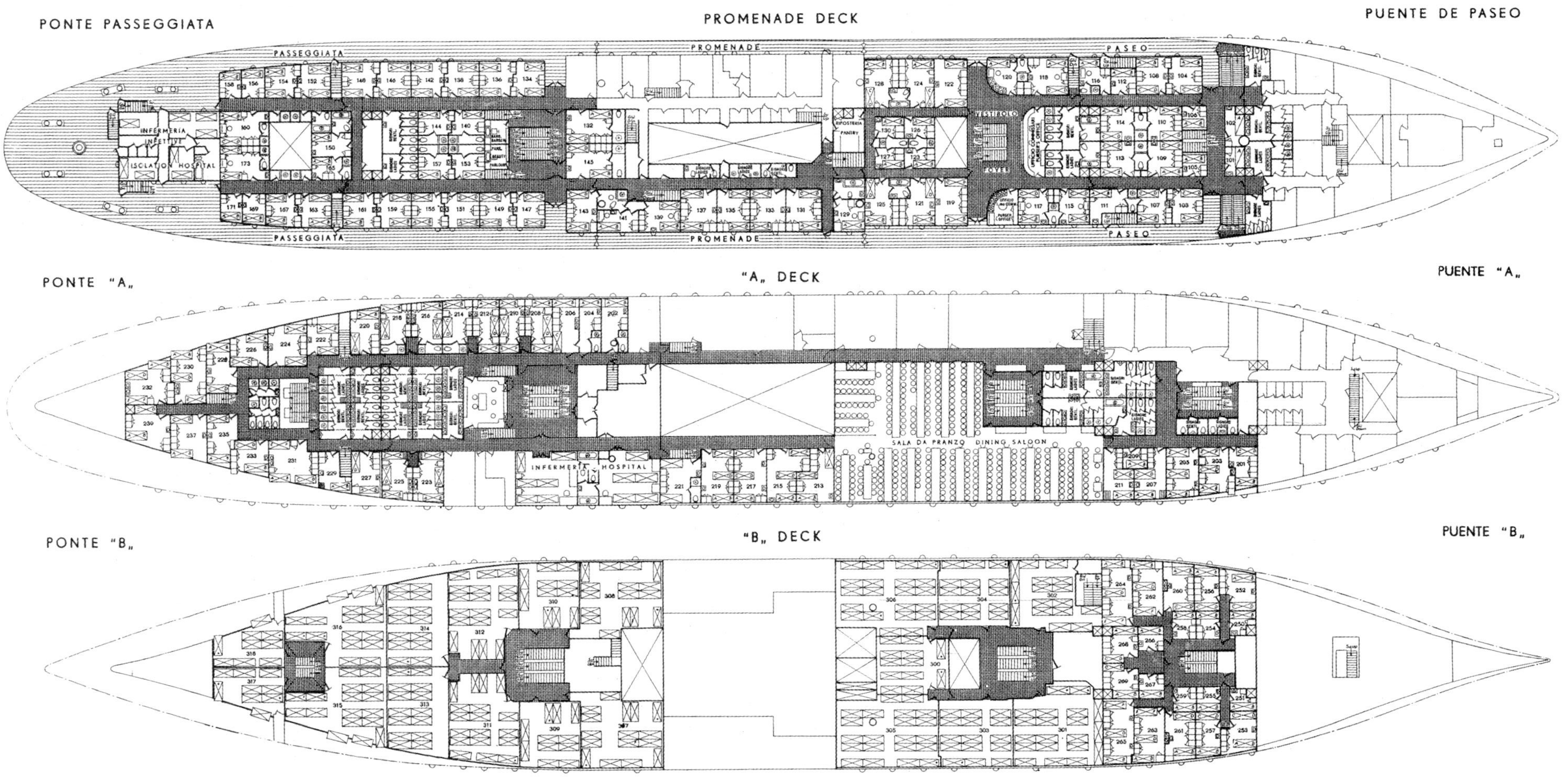

CAPACITÀ POSTI IN CABINE
CABIN CAPACITY

PONTE PASSEGGIATA / PROMENADE DECK

9 Cabine Esterne / Outside Cabins × 2 = 18 letti / berths
35 " " × 4 = 140 "
2 Cabine Interne / Inside Cabins × 1 = 2 "
13 " " × 2 = 26 "
8 " " × 4 = 32 "
67 Cabine con / Cabins fitted with . . . . . . **218** "

— delle quali cabine / of which cabins —

N 4 con bagno privato / with private bath = 12 posti / berths
" 18 con doccia privata / with private shower = 54 "
" 11 associabili ★ / connecting ★ . . . . = 36 "
N° **33** cabine / cabins Totale / Total = **102** "

PONTE / DECK "A"

3 Cabine Esterne / Outside Cabins × 2 = 6 letti / berths
17 " " × 4 = 68 "
14 " " × 6 = 84 "
1 Cabina Interna / Inside Cabins × 2 = 2 "
35 Cabine con / Cabins fitted with . . . . . **160** "

— delle quali cabine / of which cabins —

N° 5 con doccia privata / with private shower = 16 posti / berths

PONTE / DECK "B"

6 Cabine esterne / Outside cabins × 4 = 24 letti / berths
4 " " × 6 = 24 "
8 Cabine interne / Inside cabins × 2 = 16 "
2 " " × 4 = 8 "
20 Cabine con / Cabins fitted with . . . . . . **72** "

DISPONIBILI / SPARES

1 Appartamento (cabina-bagno-salotto) / Suite (cabin-bath-living room) × 3 = 3 posti / berths
1 Cabina esterna / Outside cabin . . . . × 2 = 2 "
2 Cabine con / Cabins fitted with . . . . **5** "

★ con Bagno o Doccia in comune con altra cabina adiacente.
★ with Bath or Shower in common with other adjacent cabin.

NOTA - I bagni e le doccie sono tutti con W.C.
NOTE - W.C. in all baths and showers.

RIEPILOGO CABINE / CABIN RECAPITULATION

Ponte Passeggiata / Promenade Deck	67	Cabine / Cabins	=	218	letti / berths
Ponte / Deck "A"	35	"	=	160	"
" "B"	20	"	=	72	"
Disponibili / Spares	2	"	=	5	"
	124 Cabine con / Cabins fitted with			**455**	"

CAPACITÀ POSTI IN DORMITORI / DORMITORY CAPACITY

PONTE / DECK "B"

Dormitorio / Dormitory		Letti / berths	di cui / of which
Dormitorio / Dormitory	300	= 24	4
"	301	= 40	4
"	302	= 32	4
"	303	= 30	2
"	304	= 30	2
"	305	= 42	2
"	306	= 42	4
"	307	= 32	4
"	308	= 32	4
"	309	= 22	2
"	310	= 22	2
"	311	= 36	2
"	312	= 36	2
"	313	= 28	2
"	314	= 28	2
"	315	= 33	4
"	316	= 33	4
"	317	= 18	4
"	318	= 18	4
Totale Total		**578**	**58**

RIEPILOGO TOTALE / TOTAL RECAPITULATION

N° **455** POSTI IN **124** CABINE DELLE QUALI **39** CON LOCALE D'IGIENE
BERTHS IN **124** CABINS OF WHICH **39** CABINS WITH SANITARY SERVICES

N° **578** POSTI IN **32** DORMITORI; DEI **578** POSTI, **58** SONO CON LETTI LARGHI (80 cm)
BERTHS IN **32** DORMITORY, OUT OF THE **578** BERTHS, **58** ARE WIDER BERTHS (32 in.)

TOTALE / TOTAL **1033** POSTI / BERTHS

EDIZIONE MAGGIO 1954

PRINTED IN ITALY - STAMPATO IN ITALIA
ARTI GRAFICHE SCHENONE - GENOVA

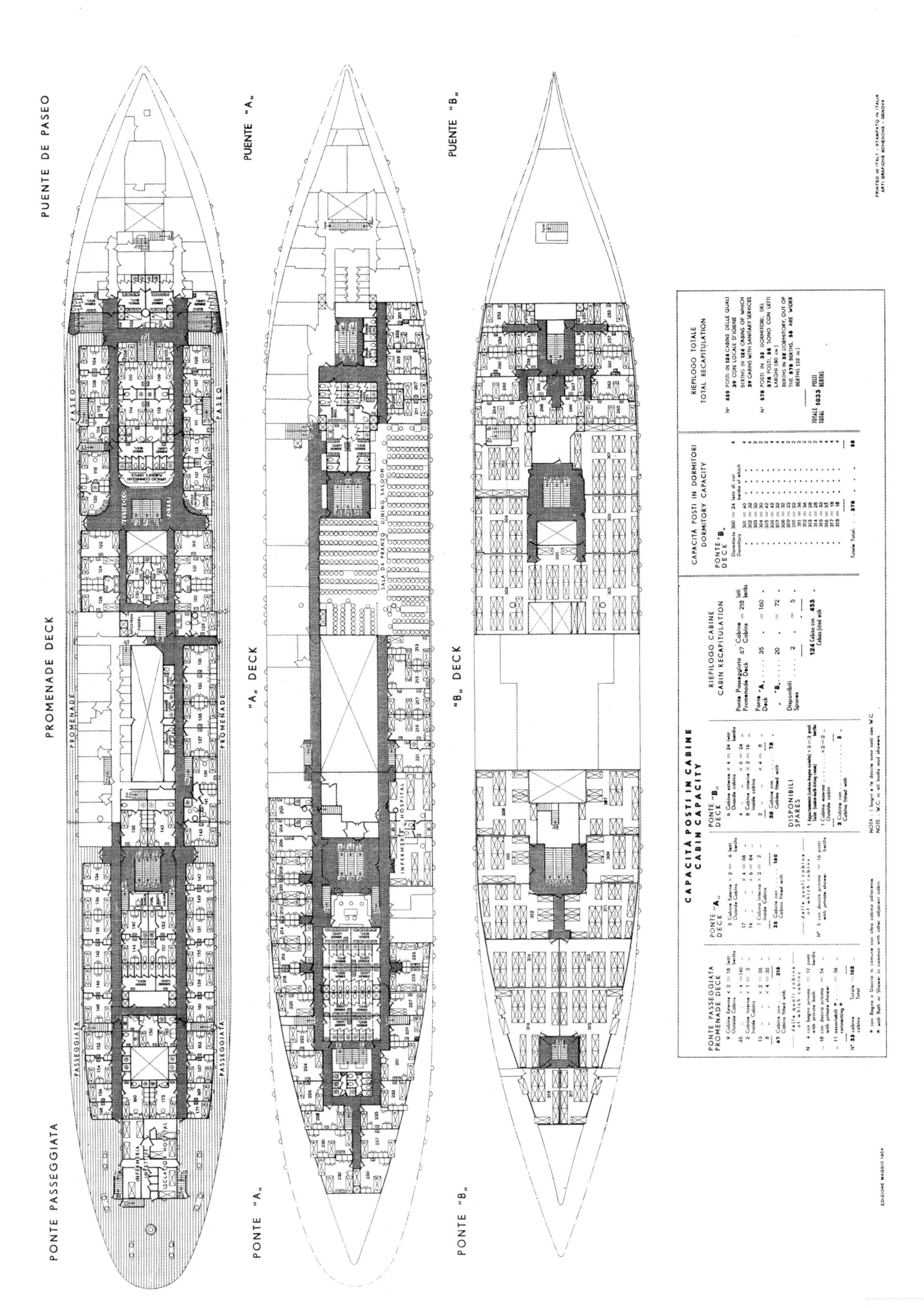

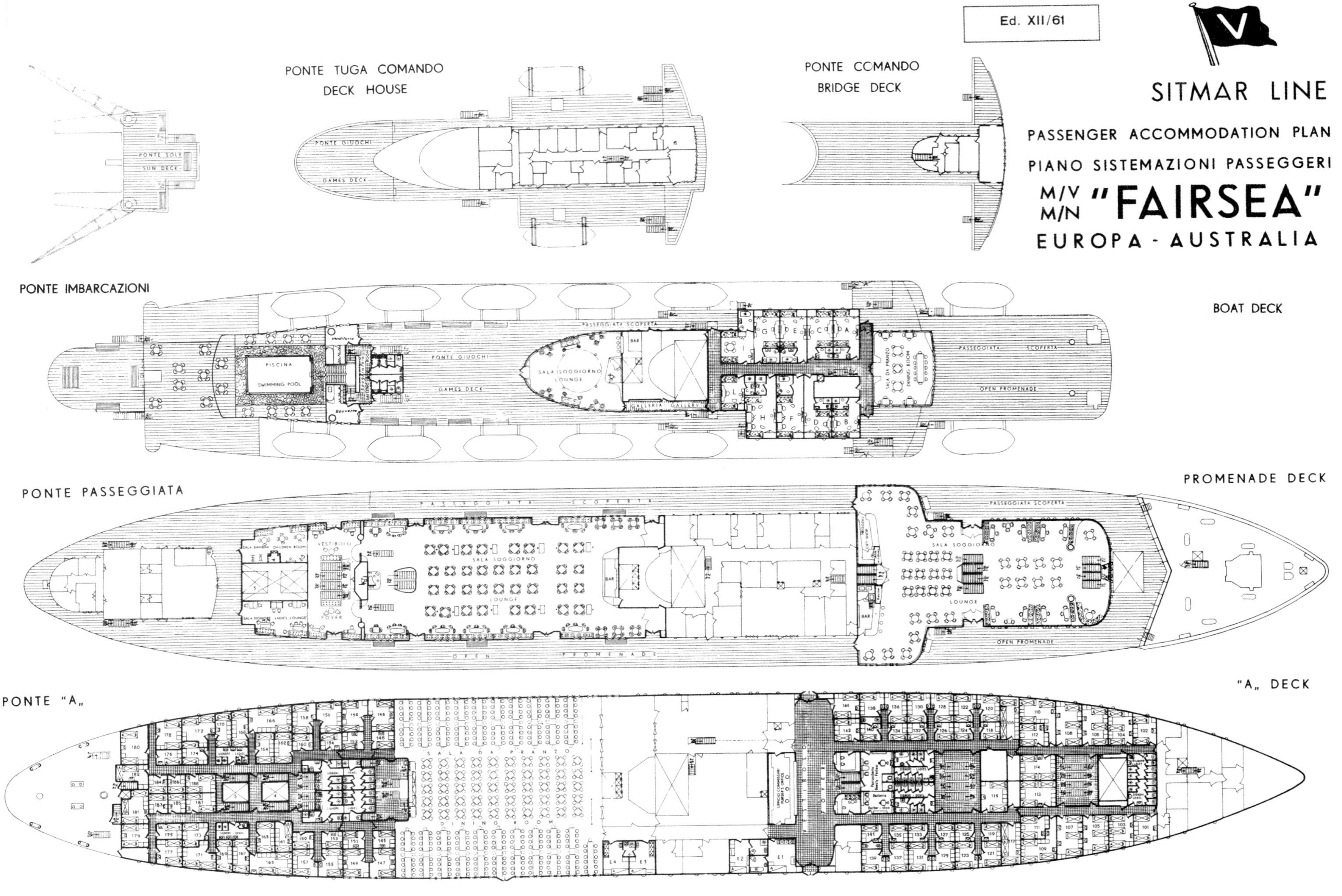

SITMAR LINE

PASSENGER ACCOMMODATION PLAN
PIANO SISTEMAZIONI PASSEGGERI

M/V "FAIRSEA"
M/N
EUROPA - AUSTRALIA

Ed. XII/61

PONTE COMANDO — BRIDGE DECK

PONTE TUGA COMANDO — DECK HOUSE

PONTE IMBARCAZIONI — BOAT DECK

PONTE PASSEGGIATA — PROMENADE DECK

PONTE "A" — "A" DECK

"B" DECK

"C" DECK

"D" DECK

PONTE "B"

PONTE "C"

PONTE "D"

LEGGENDA - GUIDE PLAN

	Spazi di ricreazione / Deck recreation space
	Passaggi e Vestiboli / Corridors and Halls
	Letti normali / Normal berths
	Letti larghi / Wide berths
	Letti bassi / Lower berths
	Letti alti / Upper berths
A . C . E	Numeri dispari / Odd numbers
B . D . F	Numeri pari / Even numbers
	Bagno / Bath
	Doccia / Shower
	W. C.
	Lavandino / Wash-Basin (WB)
	Guardaroba / Wardrobe (WR)

LEGGENDA / LEGEND

Ponte (Quarti)
Piscina
Bar
Passeggiata Infantile
Signore
Signori
Sala da Pranzo
Parrucchiere per Signore
Bar
Ufficio Commissario
Vestibolo

Deck Games
Swimming Pool
Bar
Promenade
Isolation Hospital
Ladies
Gentlemen
Dining Saloon
Hairdresser
Purser's Office
Lounge
Foyer

CAPACITÀ POSTI IN CABINE DI CLASSE TURISTICA
TOURIST CLASS CABIN CAPACITY

PONTE IMBARCAZIONI / BOAT DECK

*) 4 Cabine Esterne / Outside Cabins × 4 = 16 letti / berths
**) 6 Cabine Esterne / Outside Cabins × 4 = 24 letti / berths
10 Cabine con / Cabins fitted with = 40 letti / berths

*) con bagno e W.C. privato / with private bath and W.C.
**) con Doccia e W.C. privato / with private Shower and W.C.

PONTE "A" DECK

6 Cabine Esterne / Outside Cabins × 2 = 12 letti / berths
26 Cabine Esterne / Outside Cabins × 2 = 52 letti / berths
32 Cabine Esterne / Outside Cabins × 4 = 128 letti / berths
12 Cabine Interne / Inside Cabins × 4 = 48 letti / berths
13 Cabine Esterne / Outside Cabins × 6 = 78 letti / berths
2 Cabine Interne / Inside Cabins × 6 = 12 letti / berths
91 Cabine con / Cabins fitted with = 330 letti / berths

PONTE "B" DECK

16 Cabine Esterne / Outside Cabins × 2 = 32 letti / berths
39 Cabine Interne / Inside Cabins × 2 = 78 letti / berths
34 Cabine Esterne / Outside Cabins × 4 = 136 letti / berths
24 Cabine Interne / Inside Cabins × 4 = 128 letti / berths
6 Cabine Esterne / Outside Cabins × 6 = 36 letti / berths
2 Cabine Interne / Inside Cabins × 6 = 12 letti / berths
131 Cabine con / Cabins fitted with = 430 letti / berths

PONTE "C" DECK

19 Cabine Esterne / Outside Cabins × 2 = 38 letti / berths
34 Cabine Interne / Inside Cabins × 2 = 68 letti / berths
16 Cabine Esterne / Outside Cabins × 4 = 64 letti / berths
17 Cabine Interne / Inside Cabins × 4 = 68 letti / berths
3 Cabine Esterne / Outside Cabins × 6 = 18 letti / berths
89 Cabine con / Cabins fitted with = 256 letti / berths

PONTE "D" DECK

10 Cabine Esterne / Outside Cabins × 2 = 20 letti / berths
24 Cabine Interne / Inside Cabins × 2 = 48 letti / berths
14 Cabine Esterne / Outside Cabins × 4 = 56 letti / berths
8 Cabine Interne / Inside Cabins × 4 = 32 letti / berths
56 Cabine con / Cabins fitted with = 156 letti / berths

RIEPILOGO CABINE / SUMMARY OF CABINS

	Cabine / Cabins	letti / berths
Ponte Imbarc. / Boat Deck	10	40
Ponte "A" Deck	91	330
Ponte "B" Deck	131	430
Ponte "C" Deck	89	256
Ponte "D" Deck	56	156
Totale / Grand total	377	1212

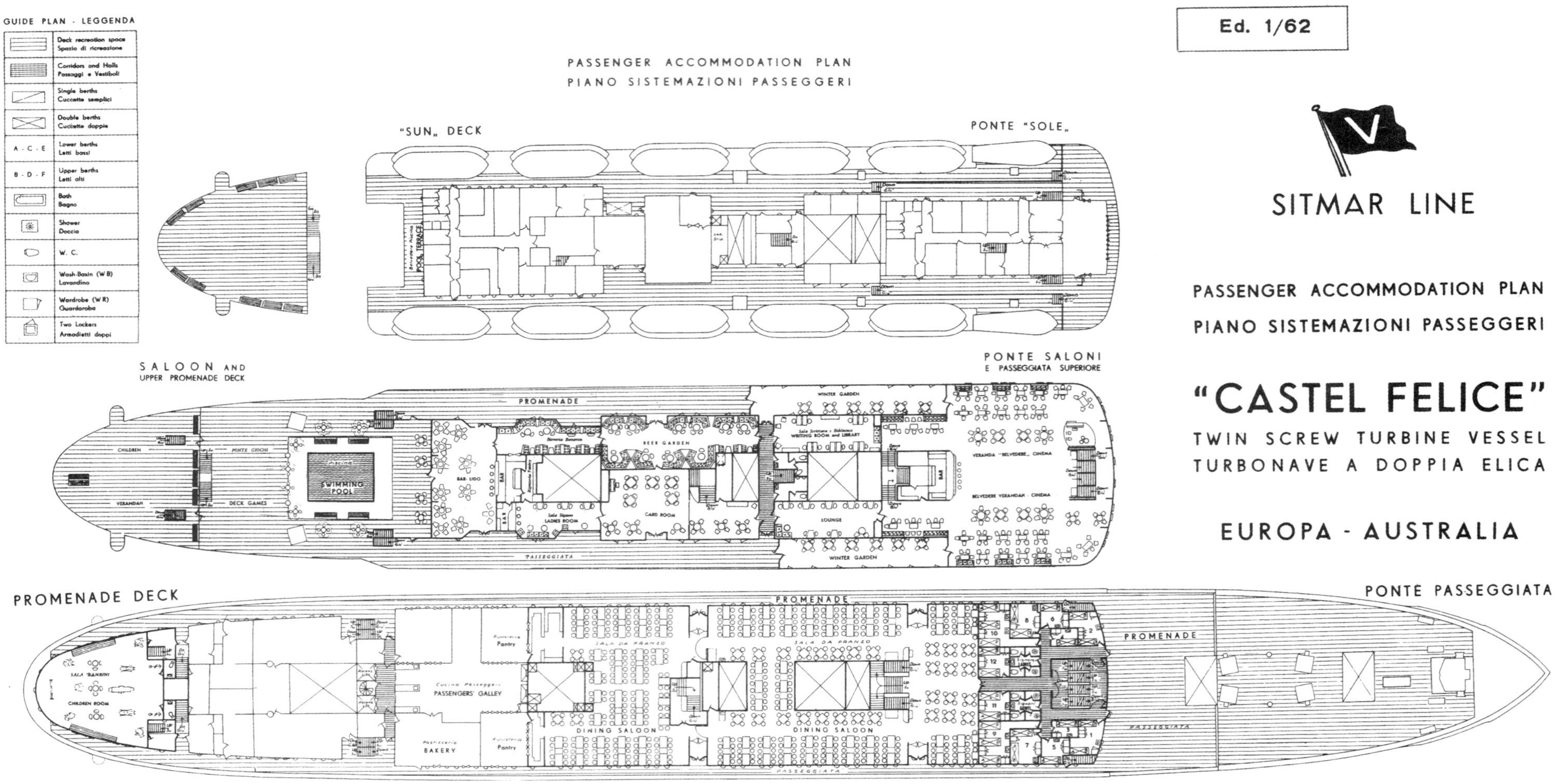
Ed. 1/62
SITMAR LINE
PASSENGER ACCOMMODATION PLAN
PIANO SISTEMAZIONI PASSEGGERI
"CASTEL FELICE"
TWIN SCREW TURBINE VESSEL
TURBONAVE A DOPPIA ELICA
EUROPA - AUSTRALIA
GUIDE PLAN - LEGGENDA
Deck recreation space
Spazio di ricreazione
Corridors and Halls
Passaggi e Vestiboli
Single berths
Cuccette semplici
Double berths
Cuccette doppie
A - C - E
Lower berths
Letti bassi
B - D - F
Upper berths
Letti alti
Bath
Bagno
Shower
Doccia
W. C.
Wash-Basin (W B)
Lavandino
Wardrobe (W R)
Guardaroba
Two Lockers
Armadietti doppi
PASSENGER ACCOMMODATION PLAN
PIANO SISTEMAZIONI PASSEGGERI
"SUN" DECK
PONTE "SOLE"
POOL TERRACE
SALOON AND
UPPER PROMENADE DECK
PONTE SALONI
E PASSEGGIATA SUPERIORE
PROMENADE
PASSEGGIATA
CHILDREN
PONTE GIOCHI
DECK GAMES
VERANDAH
SWIMMING POOL
BAR LIDO
BAR
LADIES ROOM
BEER GARDEN
CARD ROOM
WRITING ROOM and LIBRARY
LOUNGE
WINTER GARDEN
VERANDA "BELVEDERE" CINEMA
BELVEDERE VERANDAH CINEMA
PROMENADE DECK
PONTE PASSEGGIATA
SALA BAMBINI
CHILDREN ROOM
Cucina Passeggeri
PASSENGERS' GALLEY
Pantry
Pasticceria
BAKERY
SALA DA PRANZO
DINING SALOON

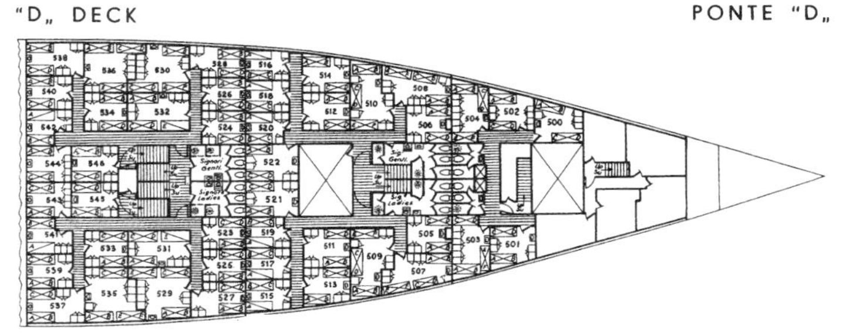

CLASSE TURISTICA - TOURIST CLASS

CAPACITÀ POSTI IN CABINE — CABIN CAPACITY

PONTE PASSEGGIATA / PROMENADE DECK

2 Cabine Esterne / Outside Cabins × 2 = 4 letti / berths
4 » » × 4 = 16 »
2 » » × 6 = 12 »
2 Cabine Interne / Inside Cabins × 2 = 4 »
2 » » × 4 = 8 »

12 Cabine con / Cabins fitted with . . . **44** letti / berths

PONTE / DECK "A"

10 Cabine Esterne / Outside Cabins × 2 = 20 letti / berths
6 » » × 3 = 18 »
41 » » × 4 = 164 »
4 » » × 5 = 20 »
6 » » × 6 = 36 »
34 Cabine Interne / Inside Cabins × 2 = 68 »
11 » » × 4 = 44 »

112 Cabine con / Cabins fitted with . . . **370** letti / berths

— delle quali cabine / of which cabins —
Nº 5 con bagno e WC privato / with bath and WC private = 18 posti / berths
Nº 2 con doccia e WC privato = 6 »
Nº 2 associabili * / connecting * = 4 »

PONTE / DECK "B"

7 Cabine Esterne / Outside Cabins × 2 = 14 letti / berths
2 » » × 3 = 6 »
43 » » × 4 = 172 »
5 » » × 5 = 25 »
3 » » × 6 = 18 »
25 Cabine Interne / Inside Cabins × 2 = 50 »
29 » » × 4 = 116 »
1 » » × 5 = 5 »
1 » » × 6 = 6 »

116 Cabine con / Cabins fitted with . . . **412** letti / berths

PONTE / DECK "C"

4 Cabine Esterne / Outside Cabins × 3 = 12 letti / berths
18 » » × 4 = 72 »
6 » » × 5 = 30 »
4 » » × 6 = 24 »
4 Cabine Interne / Inside Cabins × 1 = 4 »
2 » » × 2 = 4 »
6 » » × 3 = 18 »
14 » » × 4 = 56 »
8 » » × 5 = 40 »
2 » » × 6 = 12 »

68 Cabine con / Cabins fitted with . . . **272** letti / berths

PONTE / DECK "D"

8 Cabine / Cabins . . × 3 = 24 letti / berths
27 » . . × 4 = 108 »
2 » . . × 5 = 10 »
10 » . . × 6 = 60 »

47 Cabine con / Cabins fitted with . . . **202** letti / berths

RIEPILOGO / RECAPITULATION

Ponte Pass.te / Promen. Deck	12 Cabine con / Cabins with	44 letti / berths
Ponte «A» / Deck	112 » »	370 »
Ponte «B» / Deck	116 » »	412 »
Ponte «C» / Deck	68 » »	272 »
Ponte «D» / Deck	47 » »	202 »
Totale / Total	**355** cabine con / cabins with	**1300** letti / berths

— delle quali cabine / of which cabins —
Nº 5 con bagno privato / with private bath
Nº 2 con doccia privata
Nº 2 associabili * / connecting *

* con Doccia e W.C. in comune con altra cabina adiacente
* with Shower and W.C. in common with other adjacent cabin

CLASSE TURISTICA: N. **1300** POSTI IN **355** CABINE
TOURIST CLASS: N. **1300** BERTHS IN **355** CABINS

NOTA - I bagni e le docce sono tutti con W.C.
NOTE - W.C. in all baths and showers.

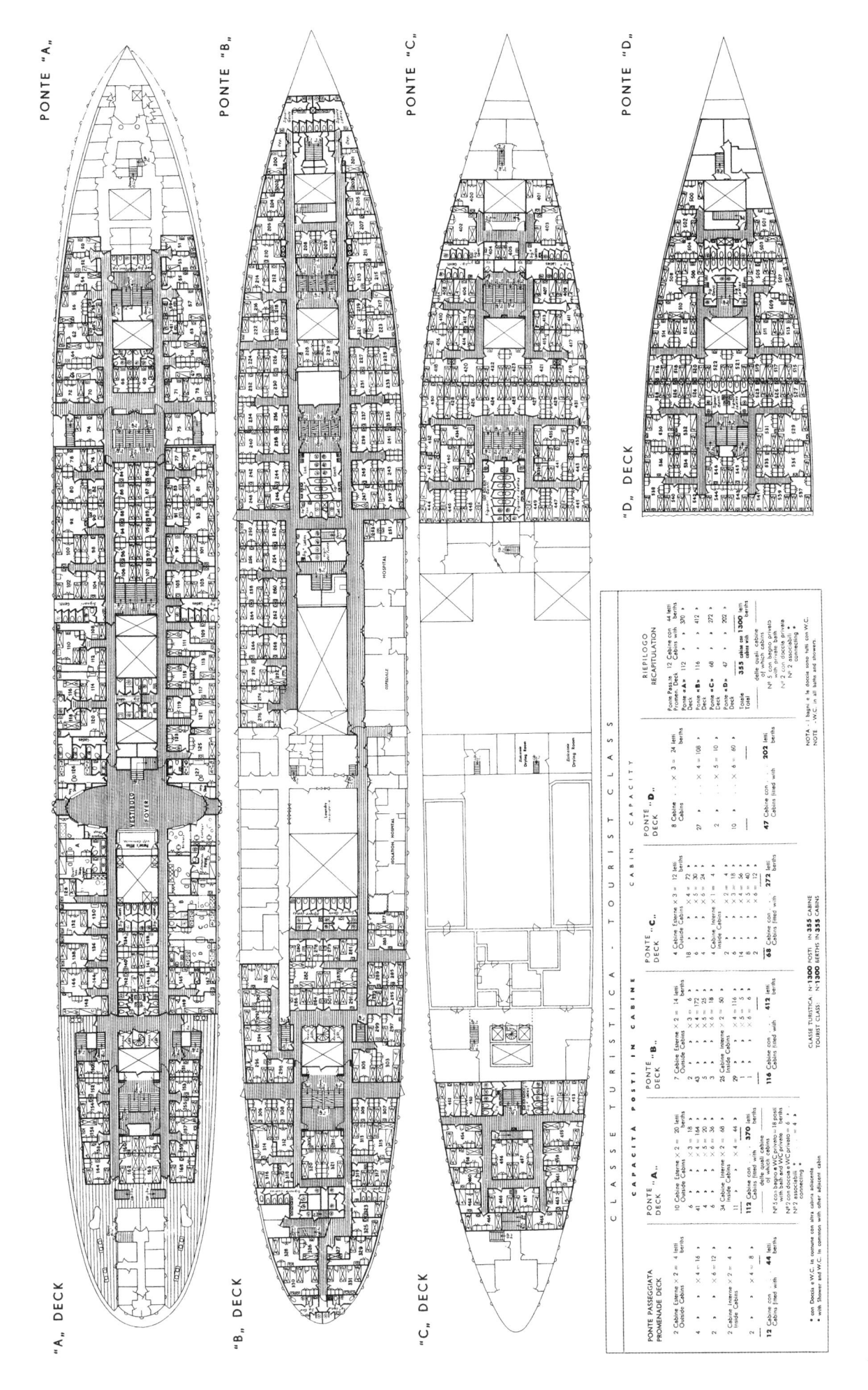

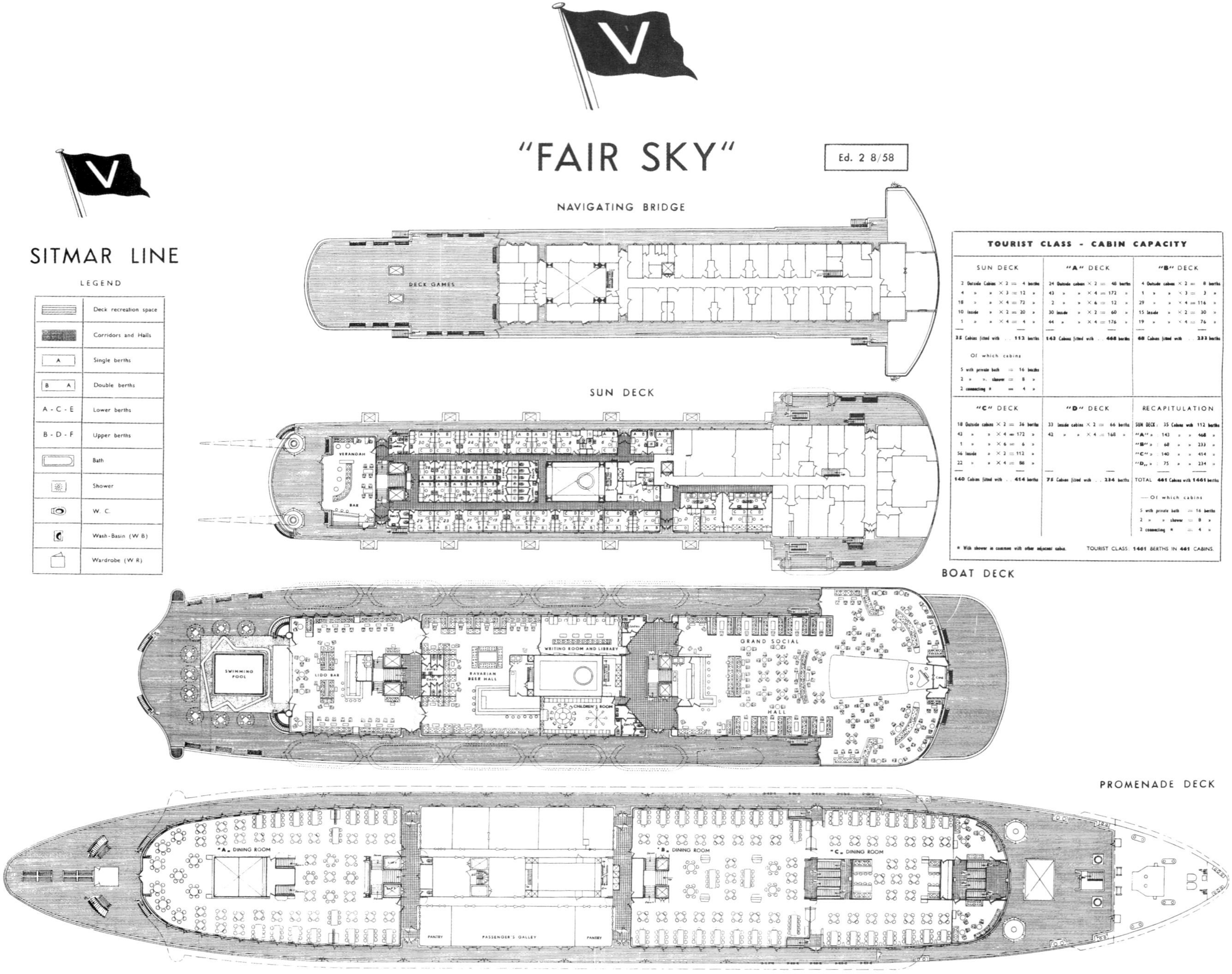

"FAIR SKY"

SITMAR LINE

Ed. 2 8/58

NAVIGATING BRIDGE
SUN DECK
BOAT DECK
PROMENADE DECK

TOURIST CLASS - CABIN CAPACITY

SUN DECK	"A" DECK	"B" DECK

(detailed cabin capacity table)

RECAPITULATION

TOTAL: 441 Cabins with 1441 berths

TOURIST CLASS 1441 BERTHS IN 441 CABINS

LEGEND
- Deck recreation space
- Corridors and Halls
- A — Single berths
- B — Double berths
- A-C-E — Lower berths
- B-D-F — Upper berths
- Bath
- Shower
- W.C.
- Wash-Basin (W B)
- Wardrobe

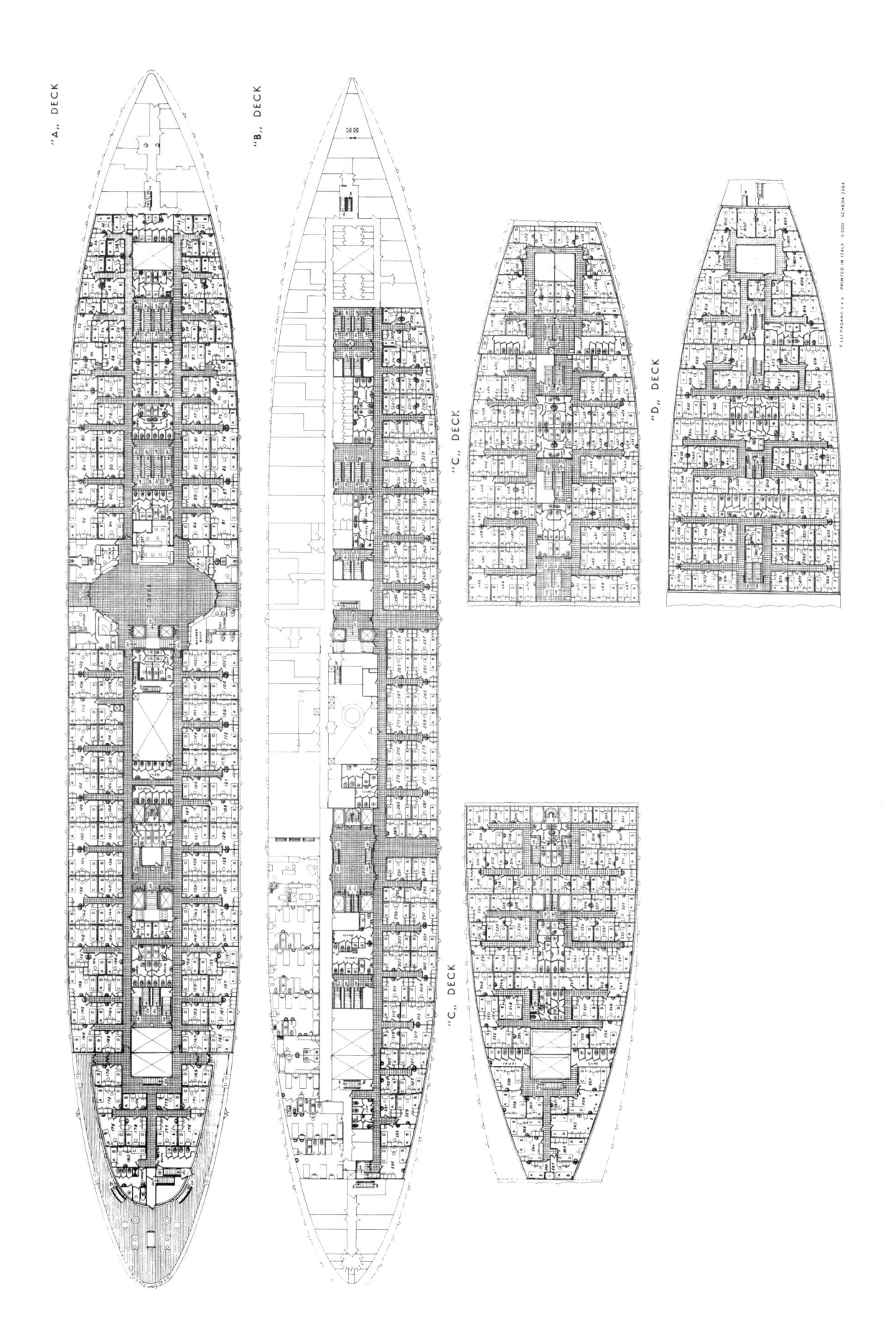

"A" DECK

"B" DECK

"C" DECK

"D" DECK

"C" DECK

FAIRSTAR Passenger Accommodation Plan

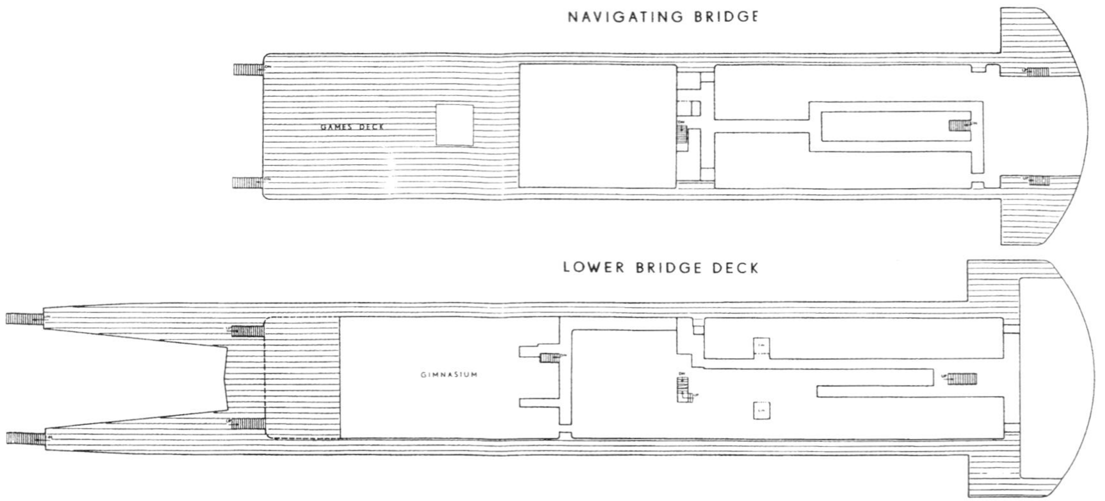

TOURIST ONE CLASS - CABIN CAPACITY

BOAT DECK
4	Outside Cabins	× 3 = 12 berths	
10	Cabins	× 4 = 40	
14	Cabins fitted with	**52 berths**	

(14 suites and cabins with private facilities)

SALOON DECK
4	Outside Cabins	× 3 = 12 berths	
8	"	× 4 = 32	
16	Inside	× 2 = 32	
6	"	× 4 = 24	
34	Cabins fitted with	**100 berths**	

of which cabins:
20 with a child berth = 20
Total **120 berths**
(32 cabins with private facilities)

"A" DECK
14	Outside Cabins	× 2 = 28 berths	
13	"	× 3 = 39	
28	"	× 4 = 112	
28	Inside	× 2 = 56	
66	"	× 4 = 264	
149	Cabins fitted with	**499 berths**	

of which cabins:
86 with a child berth = 86
Total **585 berths**
(130 cabins with private facilities)

"B" DECK
5	Outside cabins	× 2 = 10 berths	
3	"	× 3 = 9	
16	"	× 4 = 64	
21	Inside	× 2 = 42	
23	"	× 4 = 92	
68	Cabins fitted with	**217 berths**	

of which cabins:
45 with a child berth = 45
Total **242 berths**
(56 cabins with private facilities)

"C" DECK
20	Outside cabins	× 2 = 40 berths	
18	"	× 3 = 54	
20	"	× 4 = 80	
22	Inside	× 6 = 12	
34	"	× 2 = 68	
74	"	× 4 = 296	
168	Cabins fitted with	**550 berths**	

of which cabins:
69 with a child berth = 69
Total **619 berths**
(143 cabins with private facilities)

"D" DECK
6	Outside cabins	× 3 = 18 berths	
20	"	× 4 = 80	
2	Inside	× 8 = 16	
14	"	× 2 = 28	
13	"	× 4 = 52	
55	Cabins fitted with	**206 berths**	

of which cabins:
24 with a child berth = 24
Total **230 berths**
(37 cabins with private facilities)

RECAPITULATION

	Cabins with	Ad. berths plus	Ch. berths
BOAT DECK	14	52	20
SALOON DECK	34	100	20
"A" DECK	149	499	86
"B" DECK	68	217	45
"C" DECK	168	550	69
"D" DECK	55	206	24
TOTAL	488	Cabins with 1624 Ad. berths plus 244 Ch. berths

GRAND TOTAL **1868 BERTHS**

TOTAL 420 Cabins fitted with private facilities.

NAVIGATING BRIDGE

LOWER BRIDGE DECK — GAMES DECK — GYMNASIUM

BOAT DECK — JUNGLE ROOM, ZODIAC ROOM (Upper), THE SURF CLUB, SWIMMING POOL

PROMENADE DECK — THE PLAY PEN, CHILDREN SWIMMING POOL, ZODIAC ROOM (Lower), SHOP, WRITING ROOM AND LIBRARY, THE TAVERN, RAINBOW LOUNGE, AQUARIUS BAR, PASSAGE

LEGEND

	Deck recreation space
A	Single berth
A B	Double berths
A-C-E-G	Child berth
B-D-F-H	Lower berth
	Upper berth
✱	Bath
	Shower
	W.C.
	Wash-Basin
	Wardrobe

SITMAR LINE

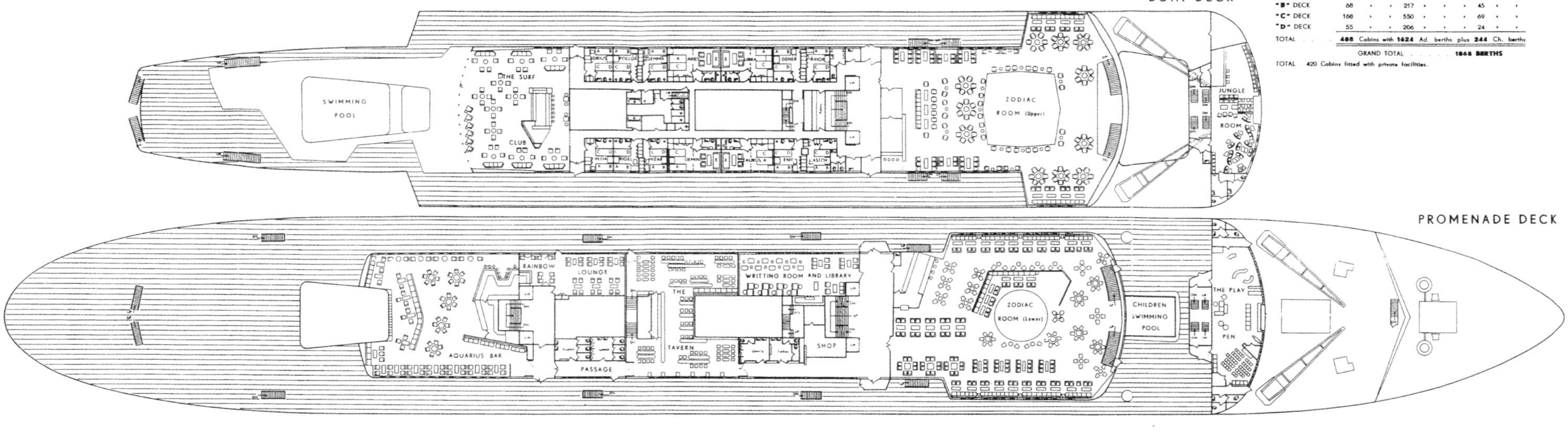

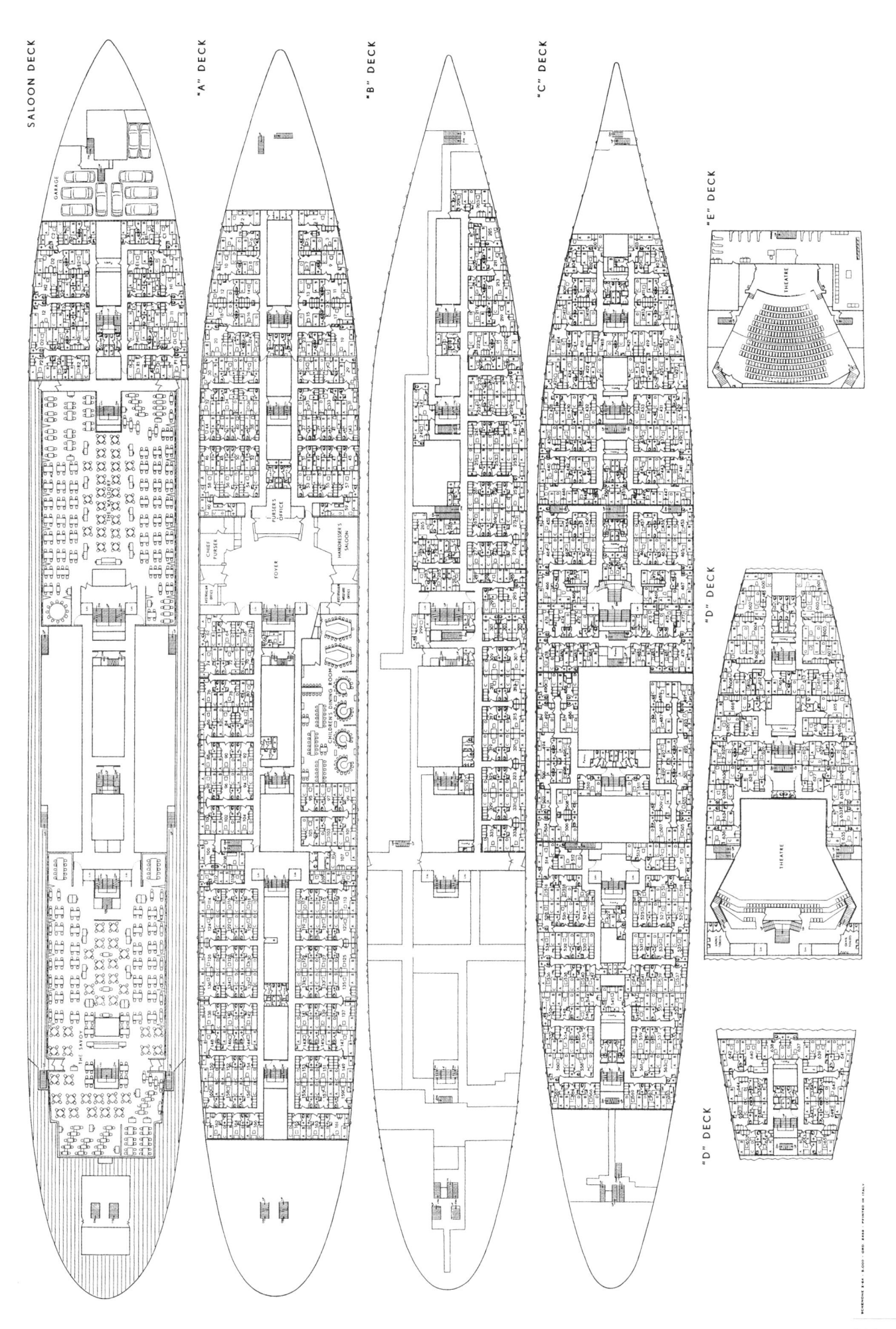

SALOON DECK "A" DECK "B" DECK "C" DECK "E" DECK "D" DECK "D" DECK

215

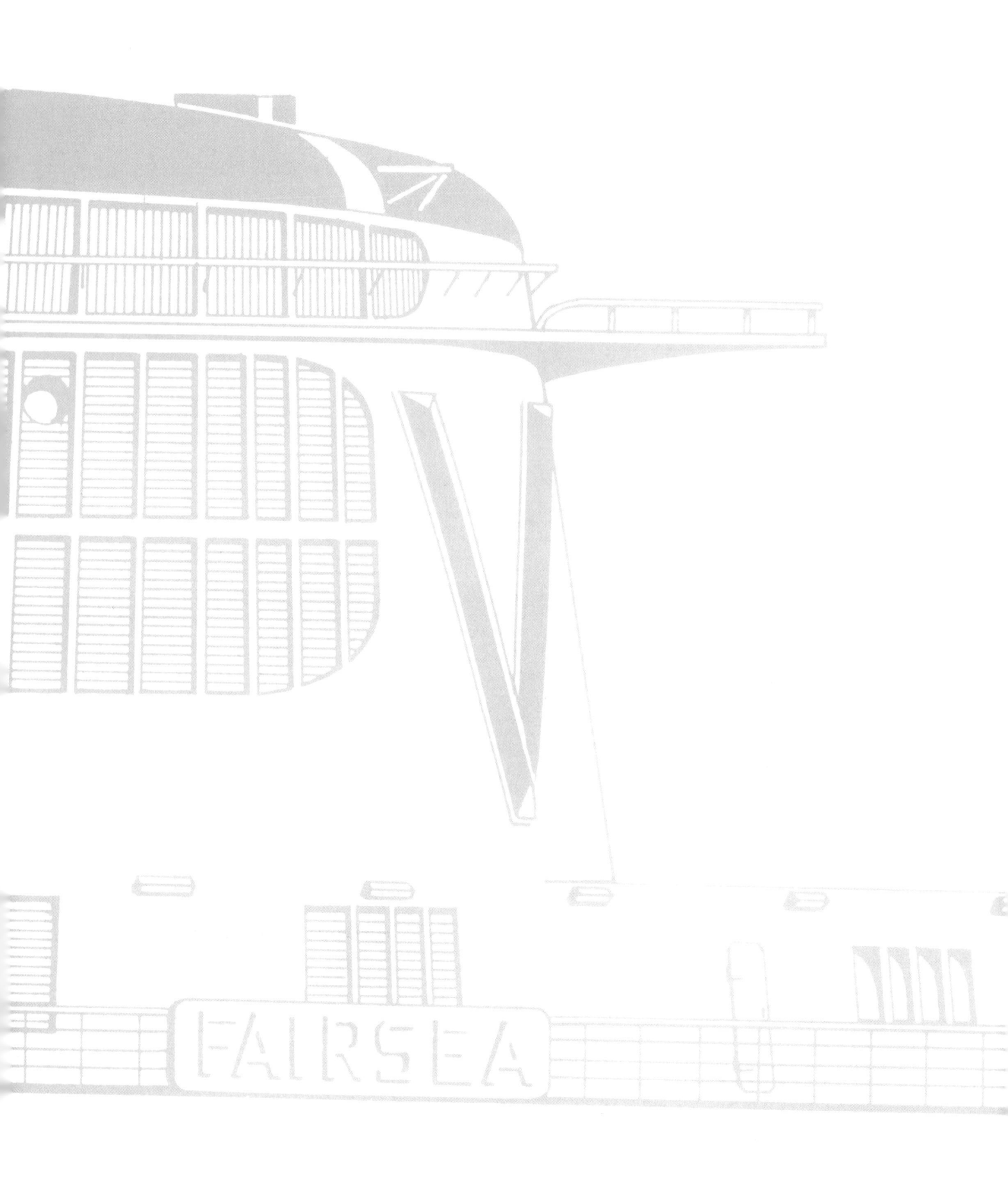

DRAWINGS

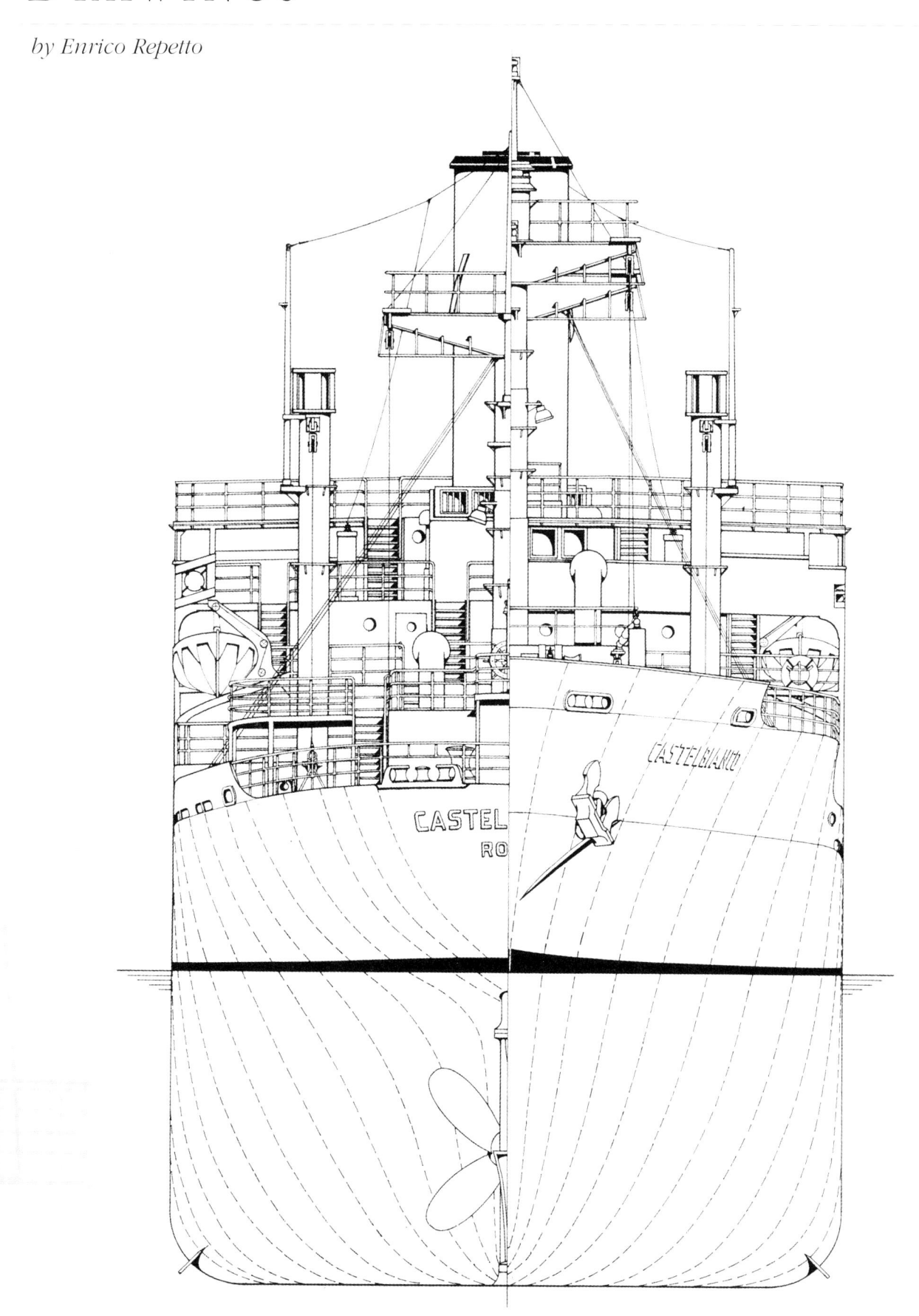

by Enrico Repetto

CASTELBIANCO (1950)

[formerly *Vassar Victory*,
then *Castel Bianco*, *Begoña*]
Builders: Bethlehem Fairfield (Baltimore)
7223 grt (3961 nrt) 10753 dwt 138.81
[133.04] x18.90 x 10.49 m
[455.2 [436.5] x 62.0 x 34.2 ft]
1 set of H.P. and L.P. DR geared turbines
6600 SHP; 17 kn
by Westinghouse Electric Co. (Pittsburg)
1132 emigrants; 122 crew

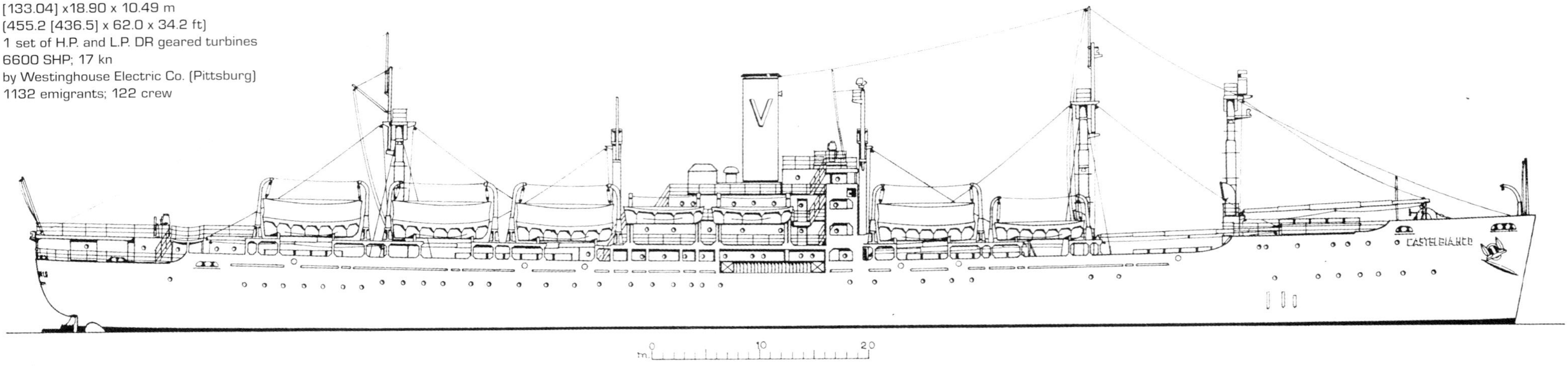

CASTEL BIANCO (1953)

[formerly *Vassar Victory*, *Castelbianco*, then *Begoña*]
Builders: Bethlehem Fairfield (Baltimore)
Re-built by C.R.D.A. (Monfalcone)
10139 grt (6686 nrt) 5747 dwt
138.81[133.04] x18.90 x 10.49 m
[455.2 [436.5] x 62.0 x 34.2 ft]
1 set of H.P. and L.P. DR geared turbines
6600 SHP; 17 kn
by Westinghouse Electric Co. (Pittsburg)
477 tourist class passengers
717 emigrants; 280 crew

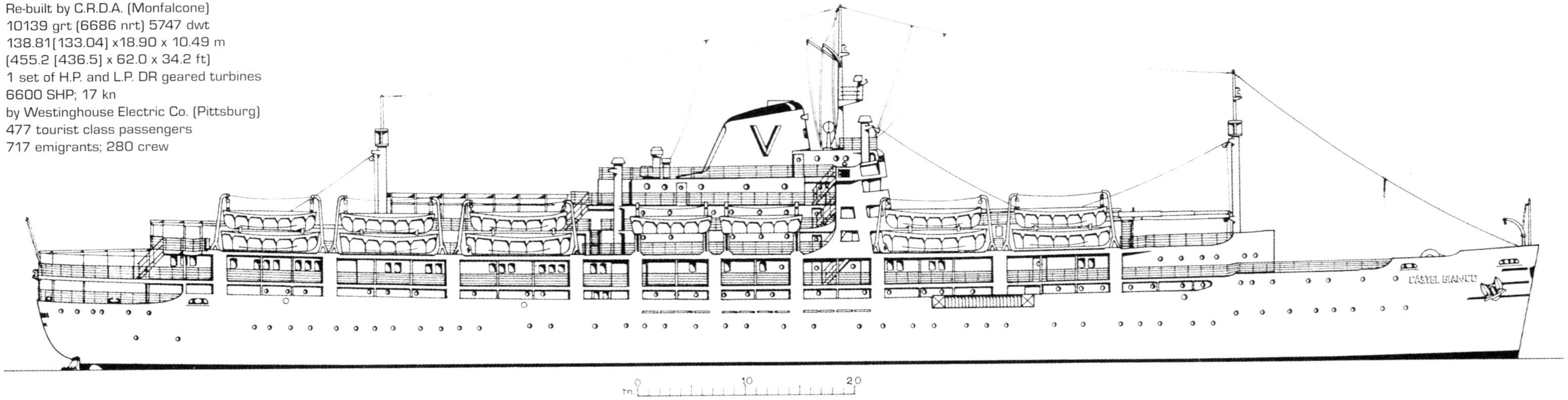

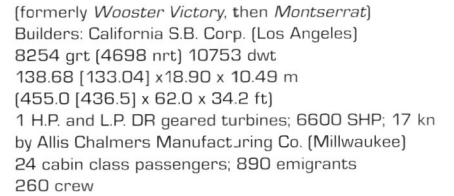

CASTEL VERDE (1950)

(formerly *Wooster Victory*, then *Montserrat*)
Builders: California S.B. Corp. (Los Angeles)
8254 grt (4698 nrt) 10753 dwt
138.68 [133.04] x 18.90 x 10.49 m
(455.0 [436.5] x 62.0 x 34.2 ft)
1 H.P. and L.P. DR geared turbines; 6600 SHP, 17 kn
by Allis Chalmers Manufacturing Co. (Millwaukee)
24 cabin class passengers; 890 emigrants
260 crew

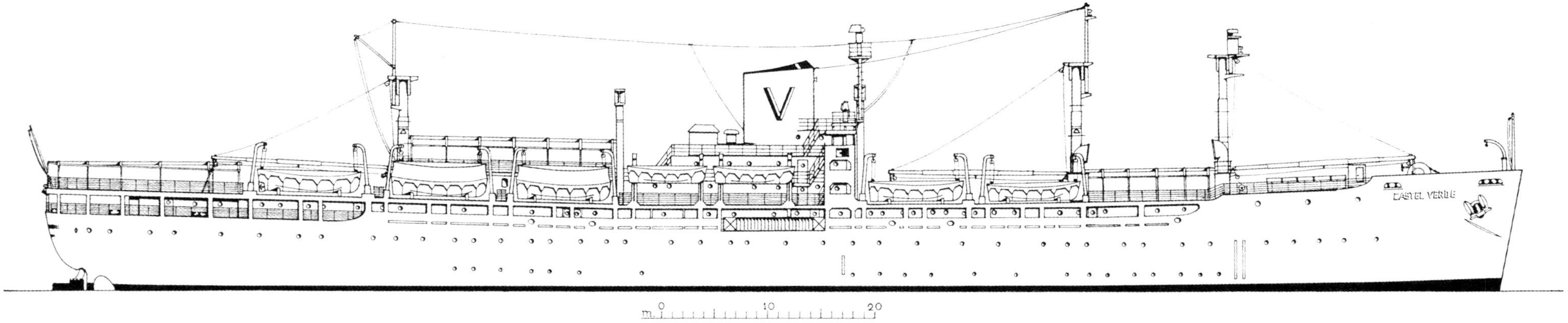

CASTEL VERDE (1953)

(ex-*Wooster Victory*, then *Montserrat*)
Builders: California S.B. Corp. (Los Angeles)
Re-built by Cantieri del Muggiano (La Spezia)
9001 grt (4758 nrt) 4375 dwt
138.68 [133.04] x 18.90 x 10.49 m
(455.0 [436.5] x 62.0 x 34.2 ft)
1 H.P. and L.P. DR geared turbines; 6600 SHP, 17 kn
by Allis Chalmers Manufacturing Co. (Millwaukee)
455 tourist class passengers; 578 emigrants; 275 crew

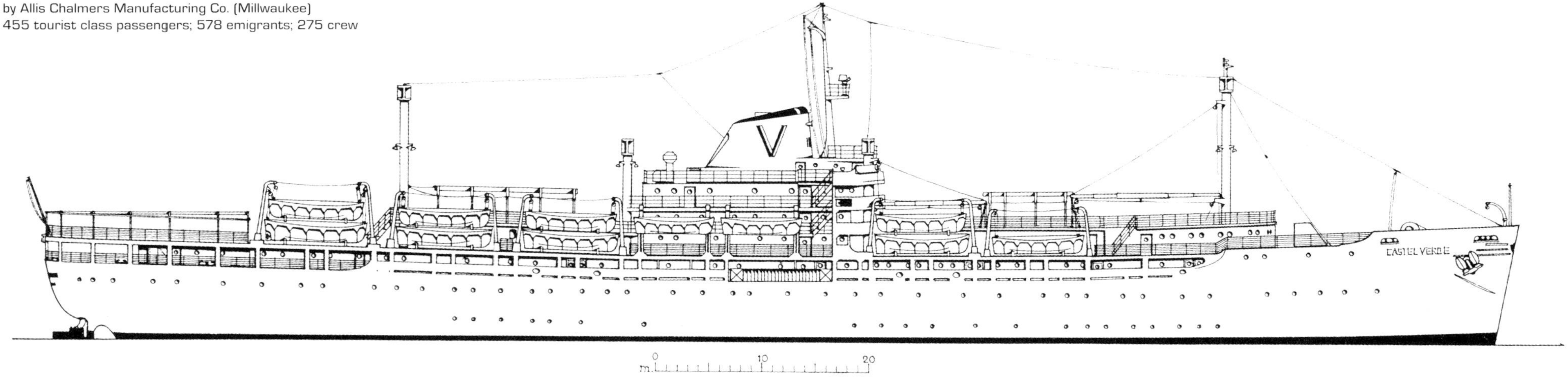

RIO DE LA PLATA (1941)

C3-P standard combi-liner
(then *HMS Charger, Fairsea*)
Builders: Sun S.B. & D.D. Co. (Chester)
11678 grt (5800 nrt) 4600 dwt
149.96 [141.57] x 21.18 x 8.32 m
(492.0 [464.5] x 69.5 x 27.1 ft)
2 single-acting 2- stroke 12-cyl. Doxford geared Diesels
8500 SHP; 16.5 kn
by builders
197 one-class passengers; 110 crew

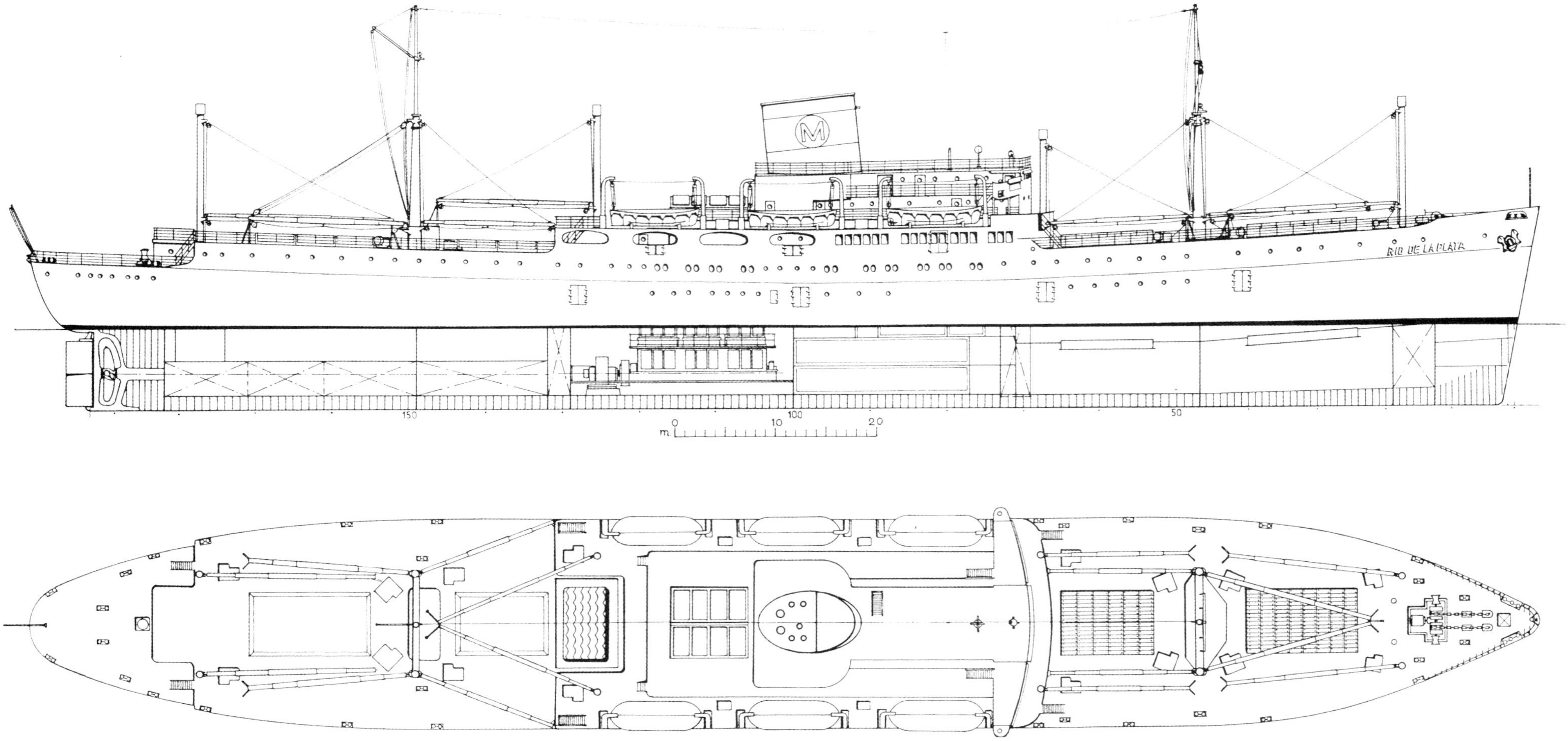

FAIRSEA (1949)
(formerly *Rio de La Plata, HMS Charger*)
Builders: Sun S.B. & D.D. Co. (Chester)
Re-built by Bethlehem Steel Corp. (Hoboken)
and Cantieri Riuniti del Tirreno (Genoa)
11678 grt (5800 nrt) 4600 dwt
149.96 [141.57] x 21.18 x 7.31 m
(492.0 [464.5] x 69.5 x 24.0 ft)
2 single-acting 2-stroke 12-cyl. Doxford geared Diesels
8500 SHP; 16.5 kn
by builders
1800 emigrants; 210 crew

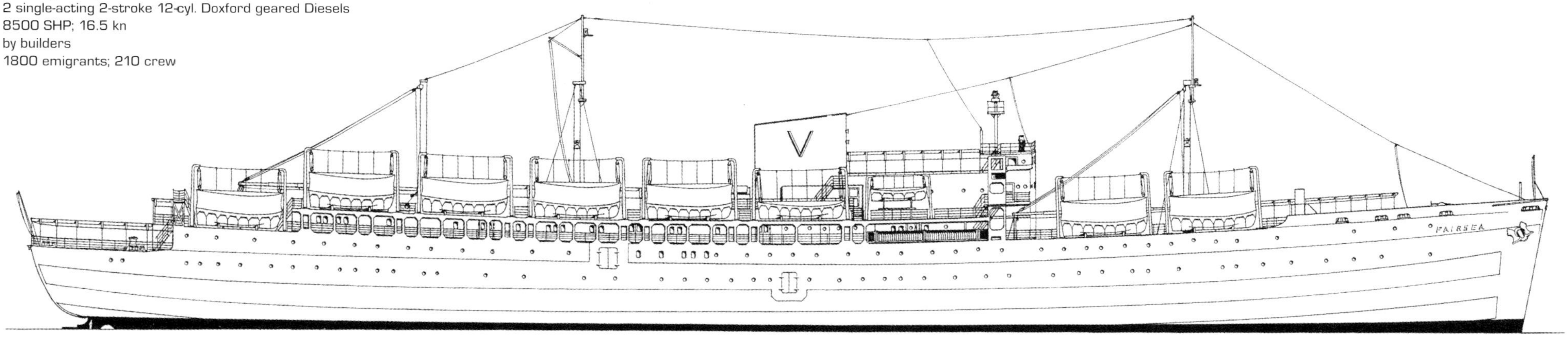

FAIRSEA (1961)
(formerly *Rio de La Plata, HMS Charger*)
Builders: Sun S.B. & D.D. Co. (Chester)
Re-built by C.R.D.A. (Monfalcone)
13433 grt (7606 nrt) 5319 dwt.
149.96 [141.57] x 21.18 x 7.31 m (492.0 [464.5] x 69.5 x 24.0 ft)
2 single-acting 2-stroke 12-cyl. Doxford geared Diesels
8500 SHP; 16.5 kn
by builders
1212 one-class passengers; 240 crew

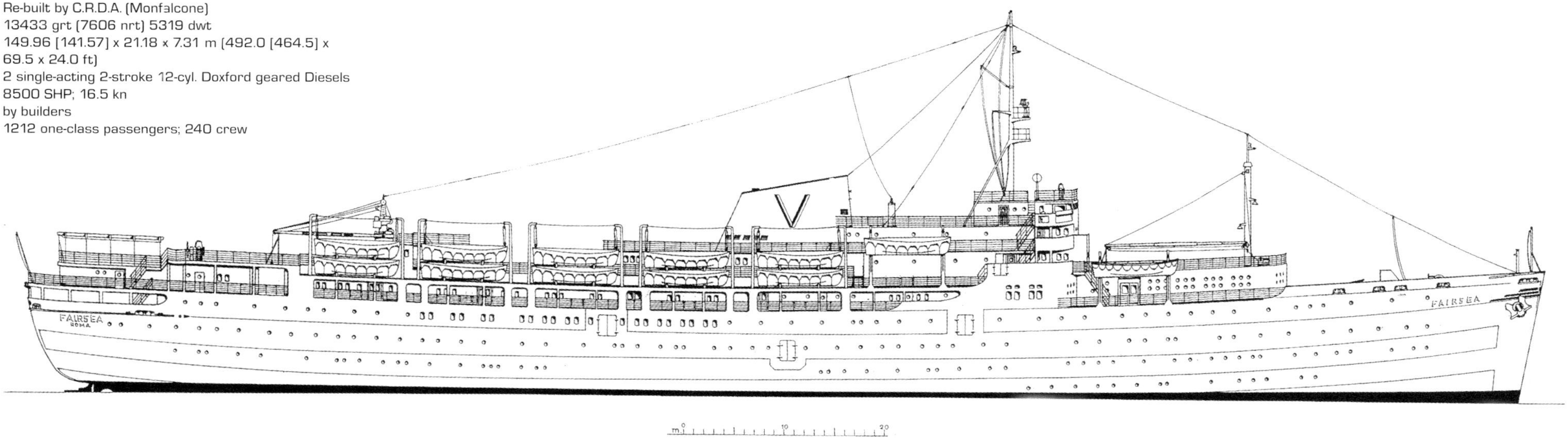

FRIESENLAND (1937)

(then *Fairsky, Castel Nevoso, Argentina Refeer*)
Builders: Howaldtswerke AG (Kiel)
5434 grt (2065 nrt) 3100 dwt
141.00 (129.00) x 16.50 x 6.04 m
(462.6 [423.2] x 54.1 x 21.9 ft)
2 9-cyl. Diesel engines; 5800 SHP, 16 kn
by M.A.N. AG (Augsburg)

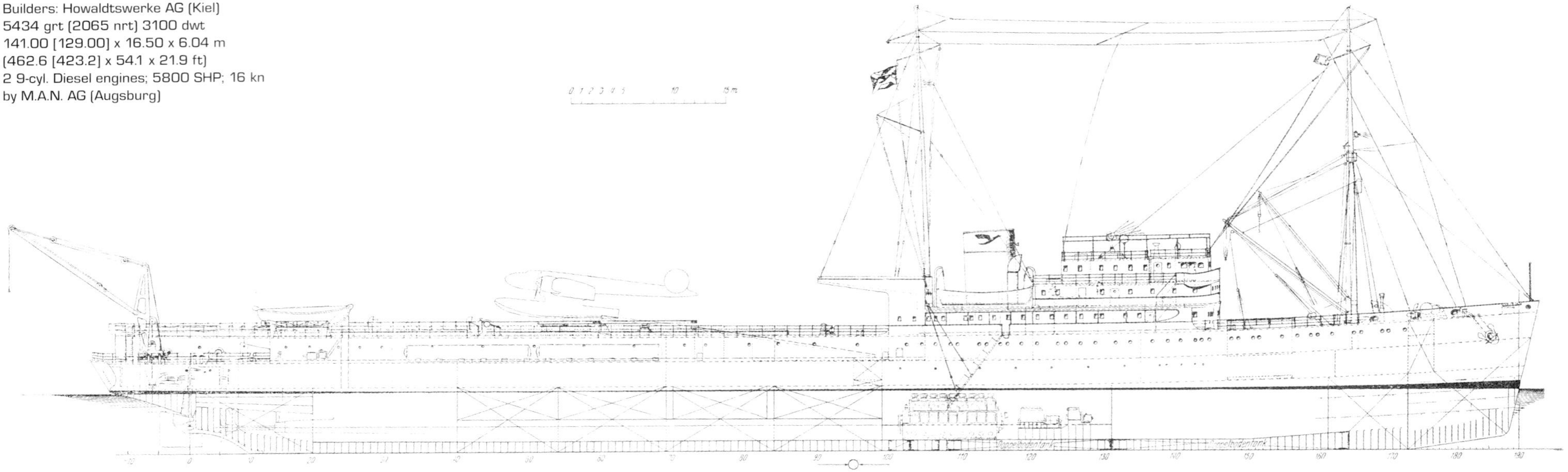

CASTEL NEVOSO (1952)

(formerly *Friesenland, Fairsky,* then *Argentina Refeer*)
Builders: Howaldtswerke AG (Kiel)
Re-built by Deutsche Werft (Hamburg)
3828 grt (1642 nrt) 2987 dwt
141.00 (129.00) x 16.50 x 5.40 m
(462.6 [423.2] x 54.1 x 27.0 ft)
2 9-cyl. Diesel engines; 5800 SHP, 16 kn
by M.A.N. AG (Augsburg)
24 passengers; 44 crew

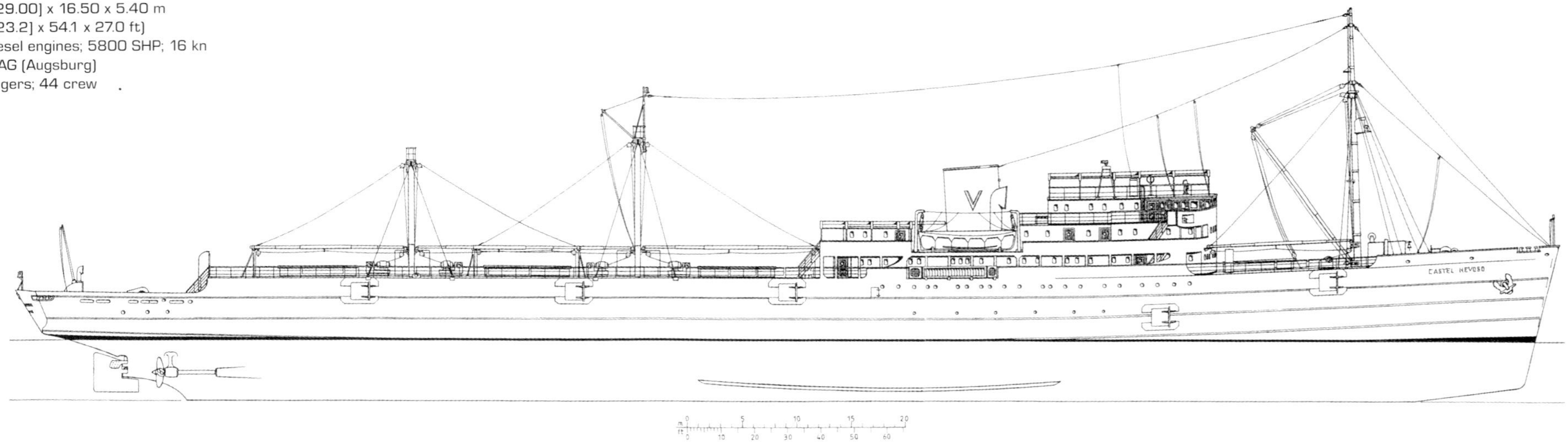

FAIRSKY (1984)

[then *Sky Princess*]
Builders: Constructions Navales et Industrielles de La Méditerranée (La Seyne)
46314 grt (22120 nrt) 7673 dwt
241.00 [203.00] x 27.80 x 7.30 m
[790.0 [665.8] x 91.1 x 24.0 ft]
2 sets of DR steam turbine; 29500 SHP; 21.8 kn
by General Electric (Lynn)
1600 passengers; 543 crew

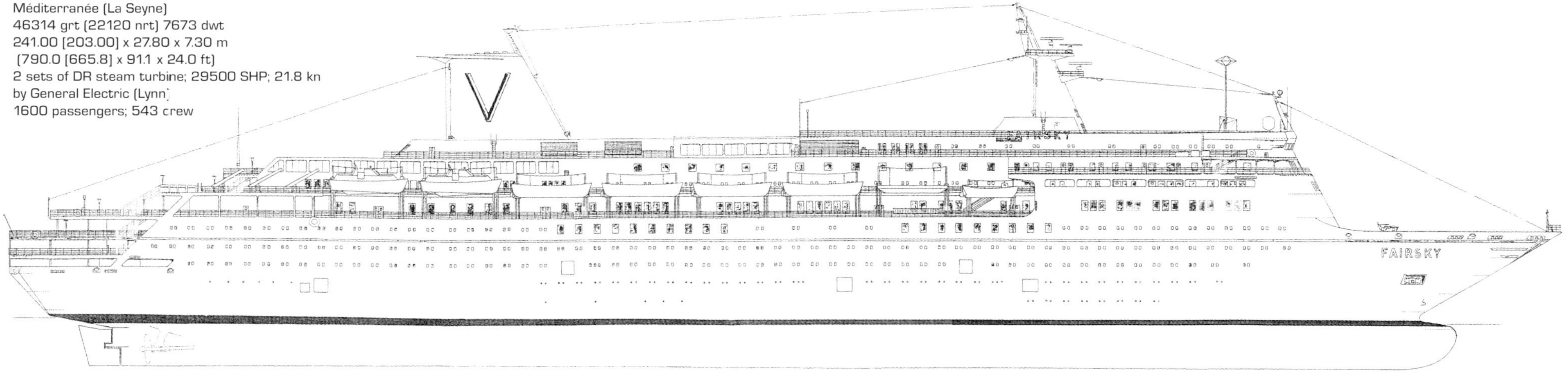

SITMAR FAIRMAJESTY (1989)

[then *Star Princess, Arcadia*]
Builders: Alsthom-Chantiers de l'Atlantique (St. Nazaire)
63524 grt (nrt) 5450 dwt
246.60 [201.00] x 32.30 x 7.70 m
[809.0 [659.4] x 106.0 x 25.3 ft]
Diesel-electric plant:
four 8-cyl. Diesel-generators, 9720kW each
2 12MW electric moto's driving FP propellers; 19.5 kn
MAN-B&W (Augsburg) - Cegelec (Belfort)
1600 passengers; 543 crew

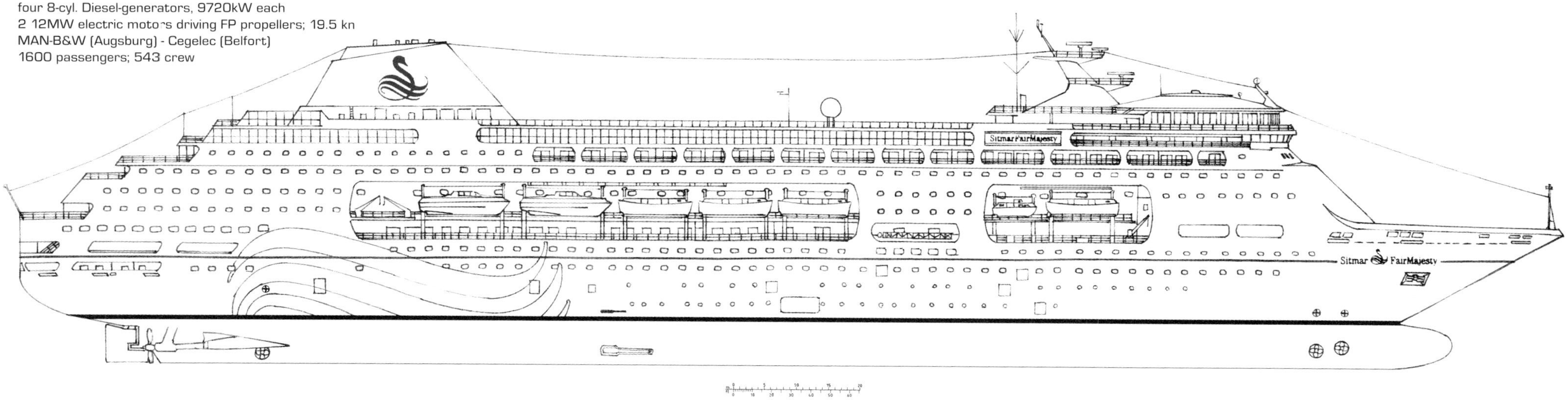

CROWN PRINCESS (1990)

Fincantieri Cantieri Navali Italiani S.p.A. (Monfalcone)
69845 grt (34907nrt) 6995 dwt
245.00 [204.4] x 32.25 x 7.80 m
[803.8 [670.6] x 105.8 ft]
Diesel-electric plant: 4 Diesel-generators, 9720kW each;
2 12MW electric motors driving FP propellers; 19.5 kn
MAN-B&W (Augsburg) - Cegelec (Belfort)
798 passengers; 656 crew

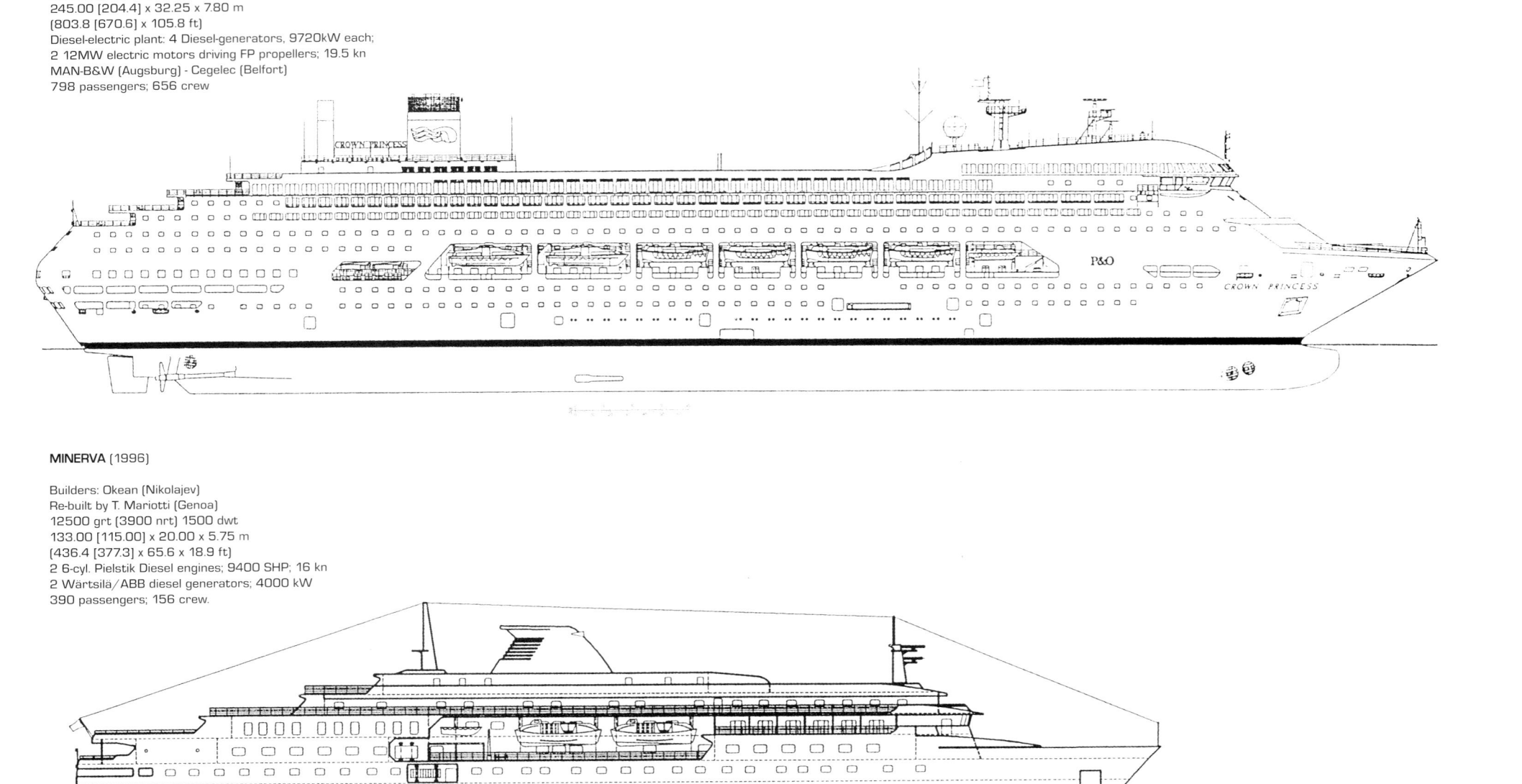

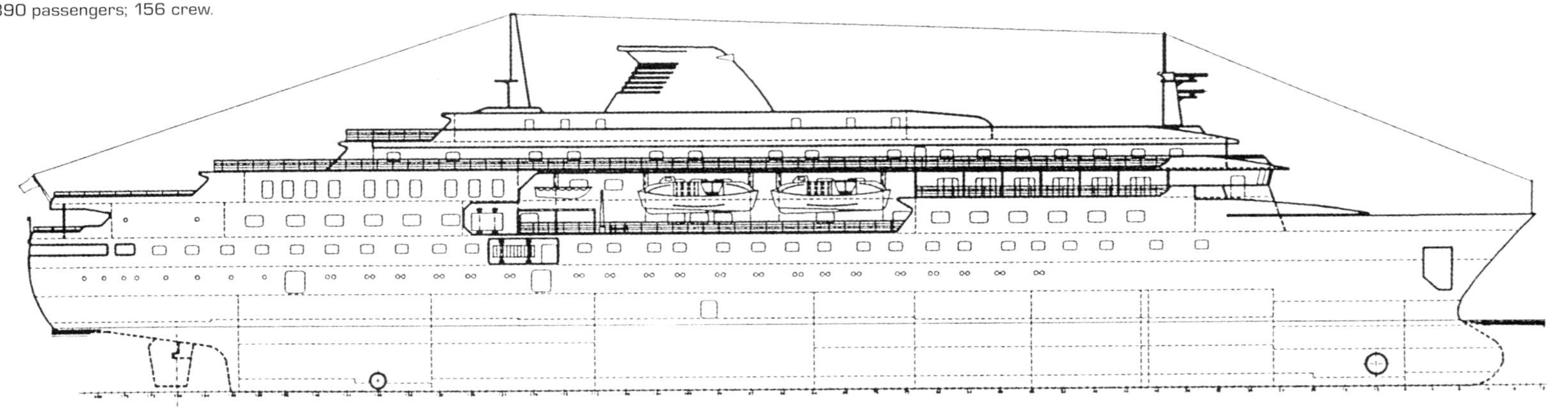

MINERVA (1996)

Builders: Okean (Nikolajev)
Re-built by T. Mariotti (Genoa)
12500 grt (3900 nrt) 1500 dwt
133.00 [115.00] x 20.00 x 5.75 m
(436.4 [377.3] x 65.6 x 18.9 ft)
2 6-cyl. Pielstik Diesel engines: 9400 SHP; 16 kn
2 Wärtsilä/ABB diesel generators; 4000 kW
390 passengers; 156 crew.

AKADEMIK NICOLAY PILYUGIN (1988)

(then *Blue Sea*, *Seven Seas Navigator*)
Builders: Admiralty Yards (St. Petersburg)
15979 grt
164.40 [150.00] x 24.00 m
[539.4 [492.1] x 78.7 ft]
4 Pielstick Diesel engines
12000 SHP; 16 kn

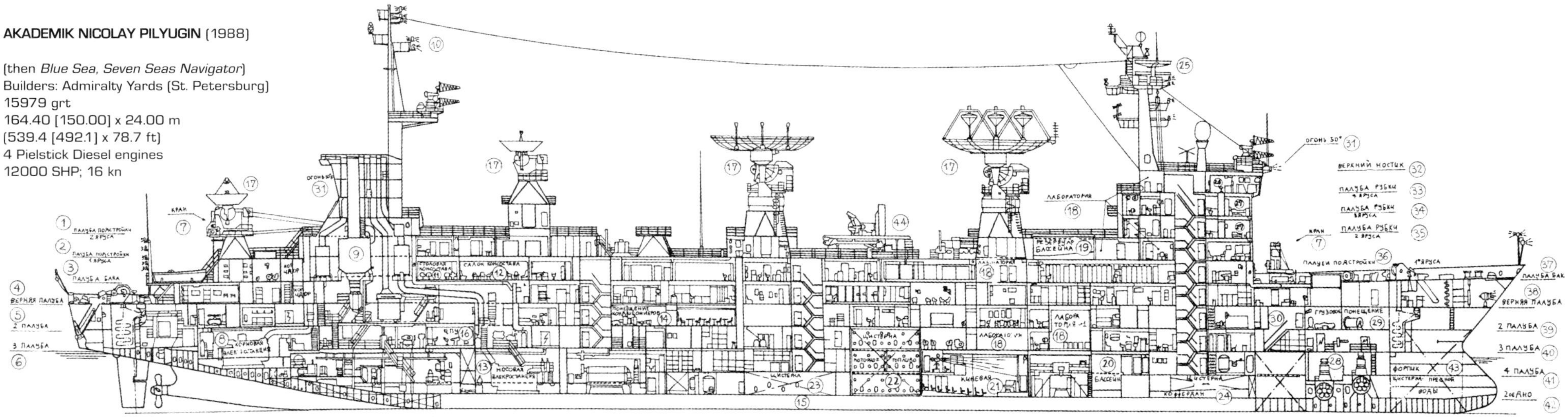

1. 2nd open deck
2. 1st open deck
3. Upper deck
4. Main deck
5. Deck 2
6. Deck 3
7. Crane
8. Aft generator room
9. Exhaust
10. Navigation lights
11. Officers' dining room
12. Officers' mess
13. Forward generator room
14. A/C station
15. Keel pipe duct
16. Control room
17. Satellite antennae
18. Laboratories
19. Open air swimming pool
20. In-door swimming pool
21. Conference room
22. Main fuel tank
23. Service fuel tank
24. Cofferdam
25. Radars
26. Bridge
27. Officers' accommodation
28. Bow thrusters
29. Spare parts store
30. Vestibule
31. 50° light
32. Bridge roof
33. 4th upper accommodation deck
34. 3rd upper accommodation deck
35. 2nd upper accommodation deck
36. 1st deck (main deck)
37. Forecastle deck
38. Main deck
39. Deck 2
40. Deck 3
41. Deck 4
42. Double bottom
43. Forepeak (fresh water tank)
44. Foldable radar

SEVEN SEAS NAVIGATOR (1999)

(formerly *Akademik Nicolay Pilyugin*, *Blue Sea*)
Builders: Admiralty Yards (St. Petersburg)
Re-built by T. Mariotti (Genoa)
25000 grt (approximate)
170.60 [150.00] x 24.80 x 6.80 m (559.7 [492.1] x 81.4 x 22.3 ft)
4 SR 8-cyl. Diesel engines; 14600 SHP; 19.5 kn by Wärtsilä (Swolle)
540 cruise passengers; 326 crew

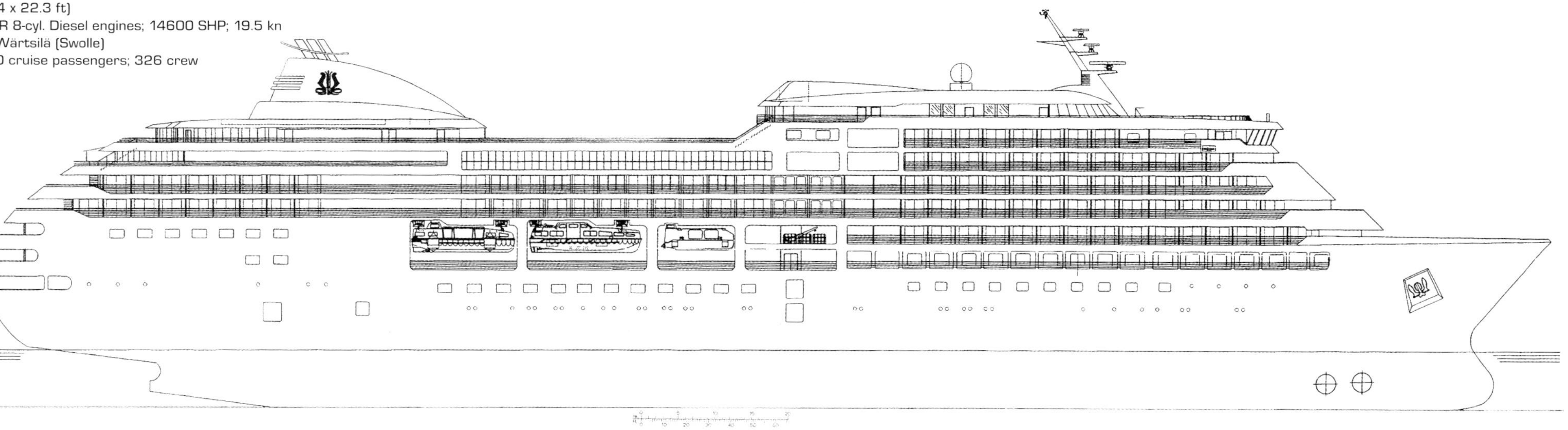

Castelbianco (1947)

(formerly *Vassar Victory*,
then *Castel Bianco*, *Begoña*)
Builders: Bethlehem Fairfield (Baltimore)
7639 grt (4571 nrt) 10753 dwt
138.81[133.04] x18.90 x 10.49 m
(455.2 [436.5] x 62.0 x 34.2 ft)
1 set of H.P. and L.P. DR geared turbines; 6600 SHP; 17 kn
by Westinghouse Electric Co. (Pittsburgh)
80 crew

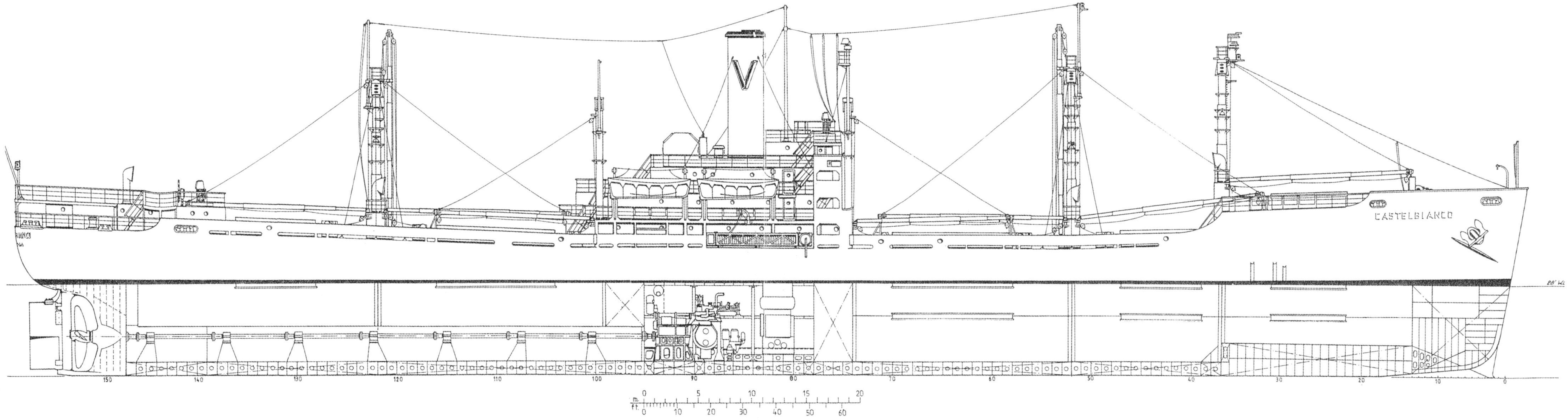

227

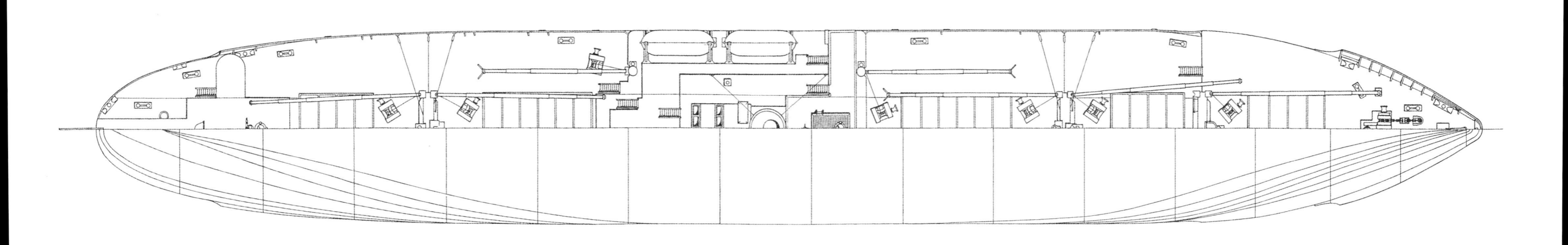

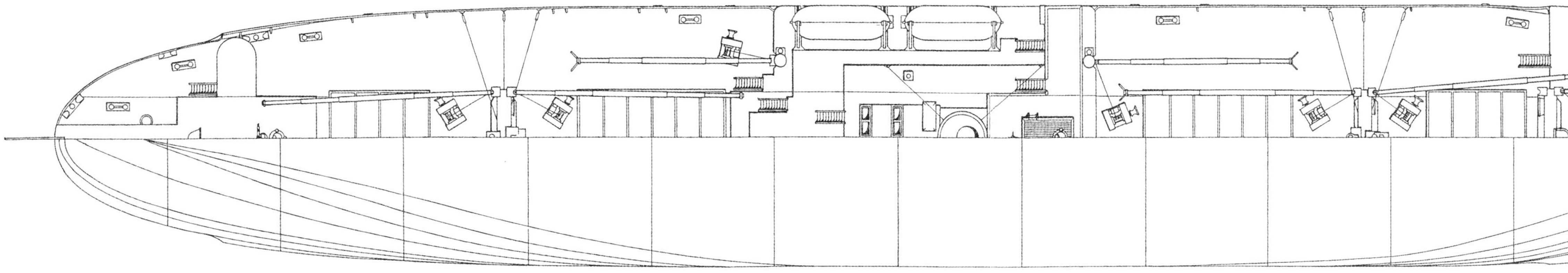

KENYA (1929)

(then *Hydra, Kenya, Keren, Fairstone, Kenya, Keren*)
Builders: Alexander Stephen & Sons Ltd (Glasgow)
9890 grt (4646 nrt) 8470 dwt
150.40 [143.25] x 19.51 x 8.96 m (493.0 [470.0] x 64.3x 29.4 ft)
2 sets of SR steam turbines; 9610 SHP; 17 kn
by builders
Passengers: 66 first class; 180 second class; 1981 deck

CASTEL FELICE (1952)

(formerly *Kenya, Hydra, Kenya, Keren, Fairstone, Kenya, Keren*)
Builders: Alexander Stephen & Sons Ltd (Glasgow)
Re-built by Cantieri Riuniti del Tirreno (Genova)
12150 grt (7140 nrt) 5210 dwt
151.80 [143.25] x 19.51 x 8.96 m (493.0 [470.0] x 64.3x 29.4 ft)
2 sets of SR steam turbines; 9610 SHP; 17 kn
by builders
596 cabin passengers; 944 emigrants

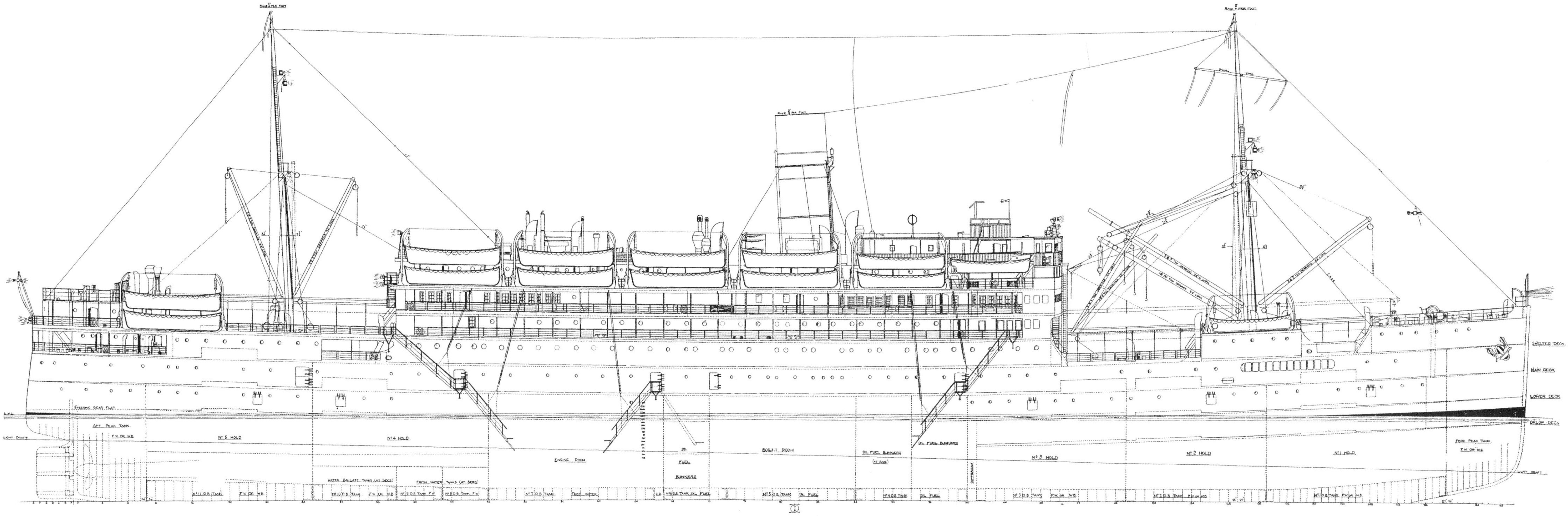
SHELTER DECK
MAIN DECK
LOWER DECK
ENGINE ROOM
BOILER ROOM
No 4 HOLD
No 5 HOLD
No 3 HOLD
No 2 HOLD
No 1 HOLD
AFT PEAK TANK
FORE PEAK TANK

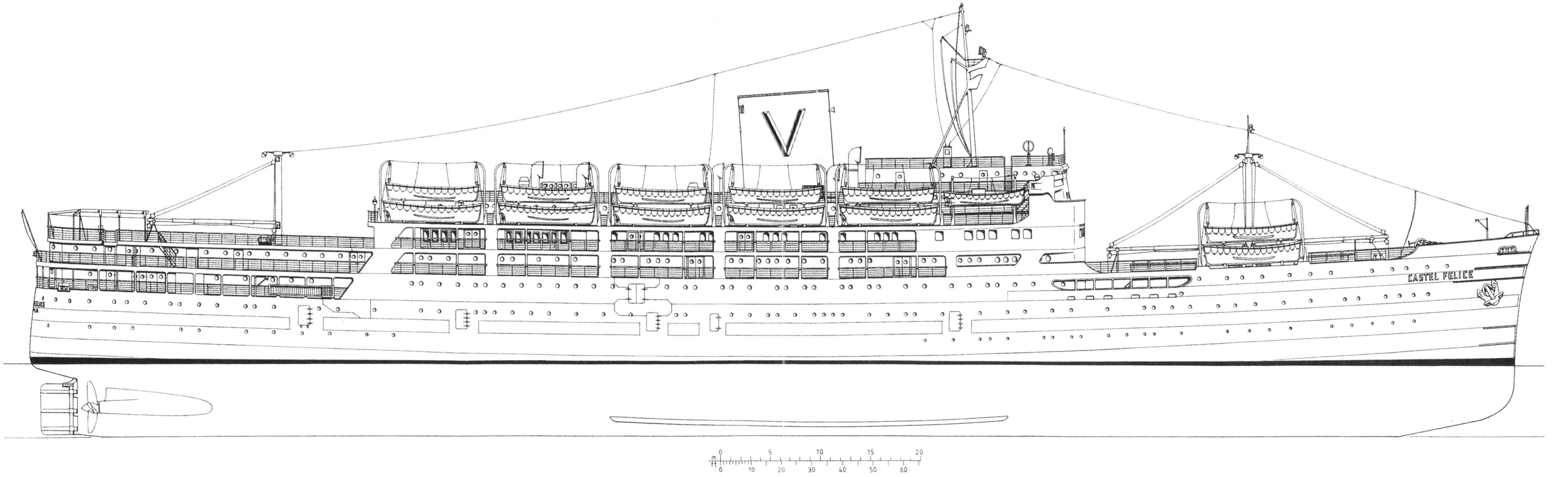
CASTEL FELICE
0
5
10
15
20
0
10
20
30
40
50
60

OXFORDSHIRE (1957)

(then *Fairstar, Ripa*)
Builders: Fairfield S.B. & Eng. Co. Ltd (Glasgow)
21619 grt (12480 nrt) 8800 dwt
185.72 [170.68] x 23.77 x 8.41
(609.3 [560.0] x 78.0 x 27.6 ft)
2 Pametrada geared turbines sets; 18000 SHP; 18.5 kn
by builders
1000 troops
passengers: 220 first class; 100 second class; 180 third class
409 crew

FAIRSTAR (1964)

(formerly *Oxfordshire*, then *Ripa*)
Builders: Fairfield S.B. & Eng. Co. Ltd (Glasgow)
Re-built by Wilton-Fijenord (Schiedam)
21619 grt (12480 nrt) 8800 dwt
185.72 [170.68] x 23.77 x 8.41
(609.3 [560.0] x 78.0 x 27.6 ft)
2 Pametrada geared turbines sets; 18000 SHP; 18.5 kn
by builders
1868 one-class passengers; 460 crew

231

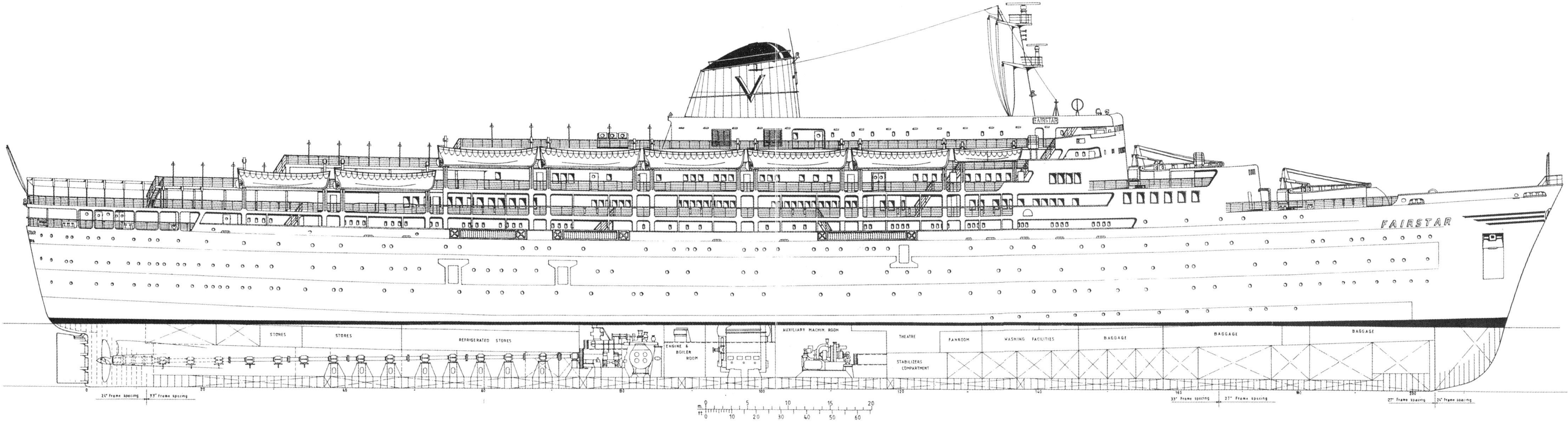

CARINTHIA (1956)

(then *Fairland*, Fairsea, *Fair Princess*)
Builders: John Brown & Co. Ltd (Glasgow)
21947 grt (11630 nrt)
185.40 [173.74] x 24.40 x 8.94 m
(608.3 [570.0] x 80.1 x 29.3 ft)
2 sets of Pametrada DR steam turbines; 24500 SHP; 20 kn
by builders
Passengers: 154 first class; 714 tourist class; 461 crew

FAIRWIND (1972)

(formerly *Sylvania*, then *Sitmar Fairwind, Dawn Princess, Albatros*)
Builders: John Brown & Co. Ltd (Glasgow)
Re-built by Arsenale Triestino San Marco (Trieste)
21985 grt (12113 nrt) 7356 dwt
185.40 [173.74] x 24.40 x 8.94 m
(608.3 [570.0] x 80.1 x 29.3 ft)
2 sets of Pametrada DR steam turbines; 24500 SHP; 20 kn
by builders
884 passengers, 470 crew

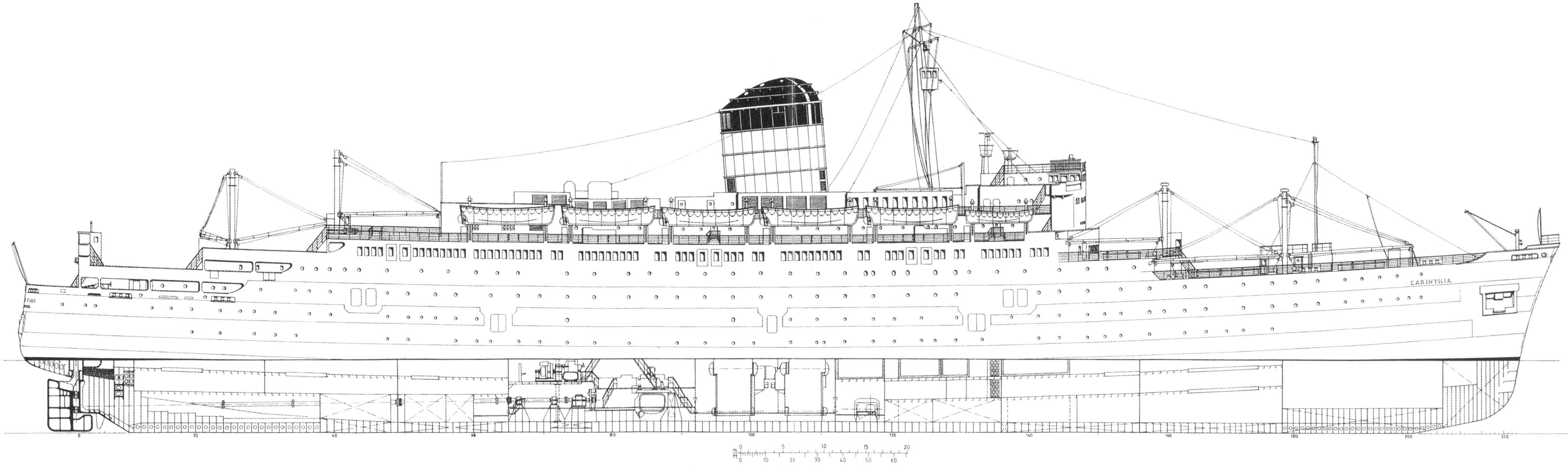
CARINTHIA

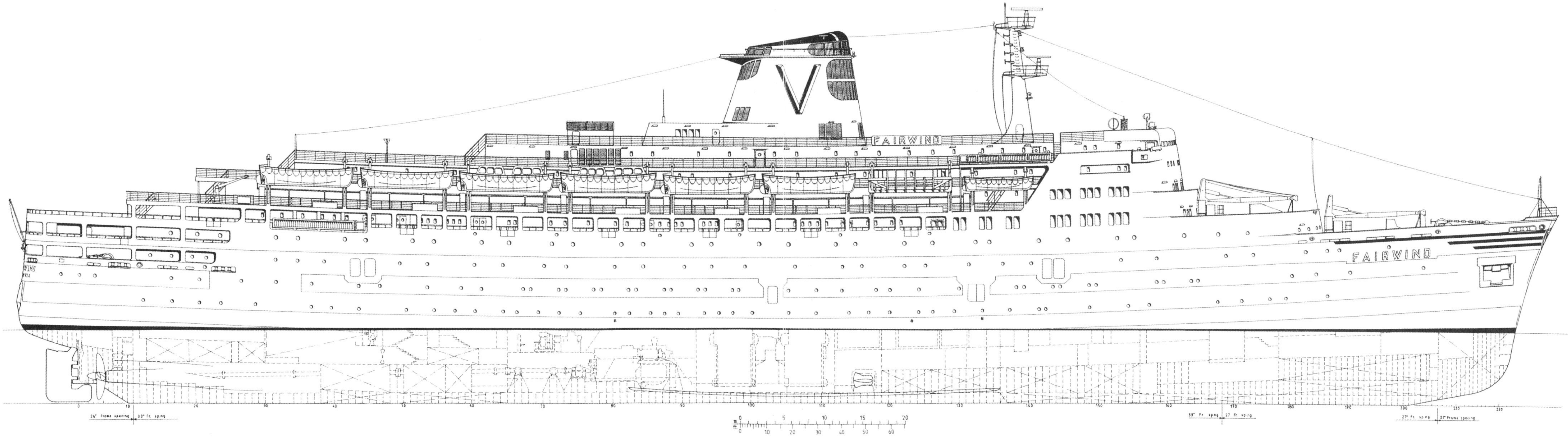
FAIRWIND
FAIRWIND

FAIRSKY (1958)

(formerly St*eel Artisan, USS Barnes, HMS Attacker, Castelforte, Castel Forte,*
then Fair Sky, Fairsky, Philippine Tourist, Fair Sky)
Builders: Western Pipe & Steel Company (San Francisco)
Re-built by T. Mariotti (Genova)
12464 grt (6682 nrt) 4950 dwt
153.00 [141.79] x 21.18 x 7.77 m
(502.0 [465.2] x 69.5 x 25.5 ft)
2 sets of DR steam turbines; 8500 SHP; 17 kn
by General Electric (Lynn)
1461 one-class passengers; 248 crew

CASTELVERDE (1938)

(formerly *Inverleith, Sunstone*)
Builders: Harland & Wolff Ltd (Belfast)
6661 grt (4093 nrt)
130.71 [125.97] x 17.00 x 10.51 m
(428.8 [410.0] x 55.8 x 34.5 ft)
1 3Exp Steam Engine; 2796 IHP; 12 kn
by builders

235

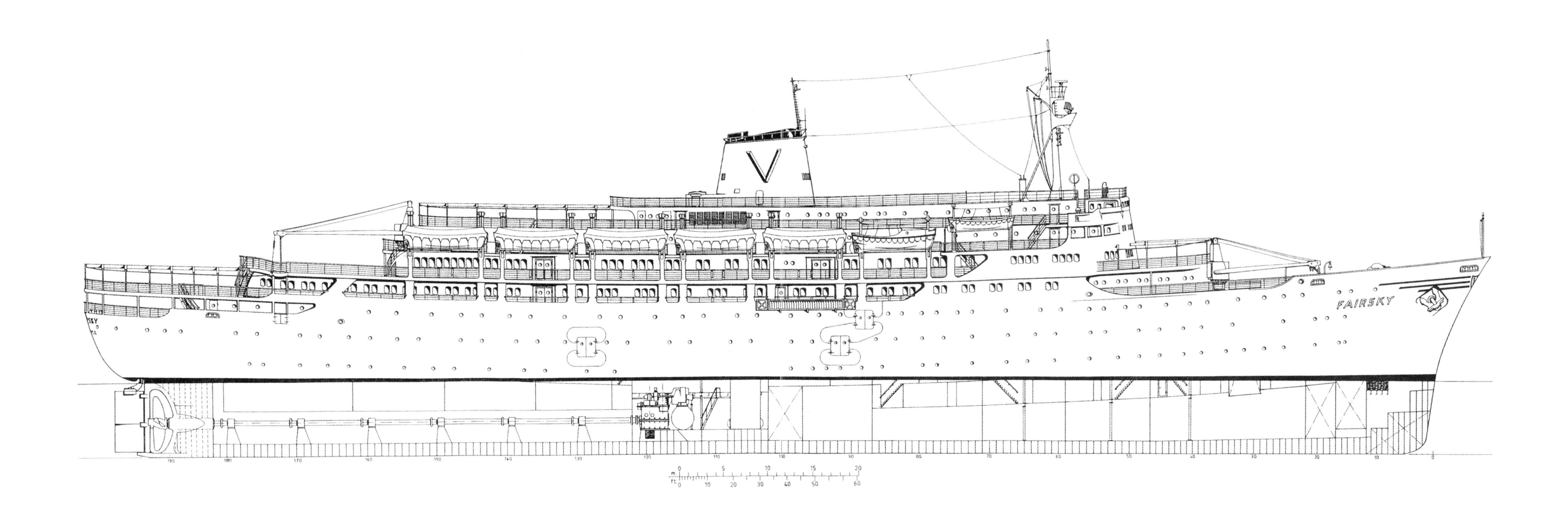

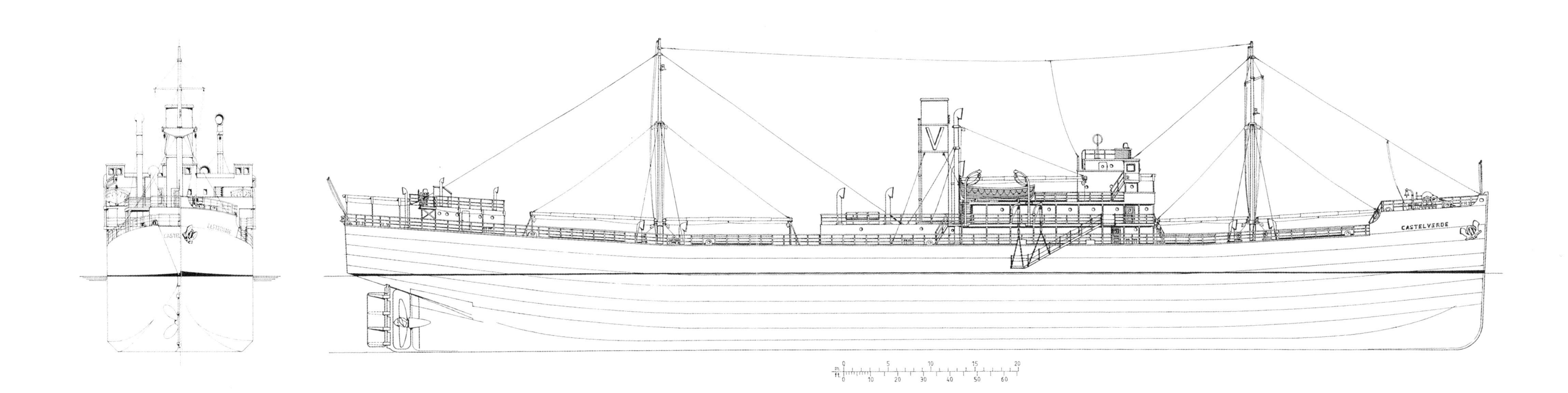

BIBLIOGRAPHY

Primary Sources

Rina, Registro Italiano Navale (Genoa), Lloyd's Register of Shipping (London), Lloyd's Voyage Records (London), Lloyd's Shipping Index (London), Lloyd's Confidential Index (London)

Newspapers and Periodicals

L'Avvisatore Marittimo (Genoa), Il Secolo XIX (Genoa), Il Corriere della Sera (Milan), Il Sole 24 Ore (Milan), Il Piccolo (Trieste), Silver Line Newsletter (London), Vlasov Group Newsletter (Monaco Monte-Carlo), V.Ships News (Monaco Monte-Carlo), Ships Monthly (Burton-on-Trent), Steamboat Bill (Chatam), The Motor Ship (London), The Shipbuilder (London), Marine Engineering (London), Transaction of the Royal Institution of Naval Architect (London), La Marina Mercantile Italiana (Genoa), Sea Breezes (Liverpool), Marine News (Kendal), G.B. Progetti (Milan)

Books

G. BADANO, *Ricordi di un Capitano*, Nuova Editrice Genovese, Genova 1992

D. J. BIBBY, *Glimpses*, Bibby Bros. & Co., Liverpool, 1991

G. BOURNEUF, *Workhorse of the Fleet*, American Bureau of Shipping, New York, 1990

A. COOKE, *Liners & Cruise Ships. some notable smaller vessels*, Carmania Press, London, 1996

A. COOKE, *Emigrant Ships*, Carmania Press, London, 1993

L. M. CORREIA, *Paquetes Portugueses*, Edições Inapa, Lisbon, 1992

L. DUNN, *Thames Shipping*, Carmania Press, London, 1996

L. DUNN, *Passenger Ships*, Adlard Coles, Southampton, 1961

G. M. FOUSTANOS, *100+7 (Liberty Ships assigned to Greece)*, Dioptra Advertising, Athens, 1997

M. GOLDBERG, *Caviar and Cargo. the C3 Passenger Ships*, Merchant Marine Museum Kings Point, New York, 1992

D. HAWS, *The Holland America Line*, TCL Publications, Newport, Uckfield, 1995

A. KLUDAS, *Great Passenger Ships of the World*, Vol. 1-6, Patrick Stephens Ltd, Wellingborough, 1986

A. KLUDAS, *Great Passenger Ships of the World Today*, Patrick Stephens Ltd, Wellingborough, 1990

P. C. KOHLER, *Sea Safari. British India S. N. Co.*, Heaton Publishing, Abergavenny, Gwent, 1995

W. H. MILLER, *Passenger Liners Italian Style*, Carmania Press, London, 1996

MITCHELL - L. SAWYER, *The Empire Ships*, Lloyd's of London Press, 1990

MITCHELL - L. SAWYER, *The Liberty Ships*, Lloyd's of London Press, 1970

MITCHELL - L. SAWYER, *British Standard Ships of World War One*, The Journal of Commerce & Shipping telegraph, Liverpool, 1968

MITCHELL - L. SAWYER, *The Victory Ships*. Lloyd's of London Press, 1980

P. PLOWMAN, *Emigrant Ships to Luxury Liners*, New South Wales University Press, Kensington, 1992

F. SERAFINI, *Ponte di Comando*, Gribaudo, Turin, 1996

I. G. STEWART, *The Ships that serve New Zealand*, 1964

I. G. STEWART, *British Tramps*, Stewart Marine Publications, Rockingham Beach, 1998

I. G. STEWART, *Liberty Ships in Peacetime*, Stewart Marine Publications, Rockingham Beach, 1992

D. WARD, *1998 Complete Guide to Cruising and Cruise Ships*, Berlitz Publishing Co., Princeton, 1998

M. H. WATSON, *Disasters at Sea*, 2nd Edition, Patrick Stephens Ltd, Sparkford, 1995

Star Princess Inaguration, P&O, London, 1989

Navi Mercantili Perdute, Ministero della Marina, Roma, 1997

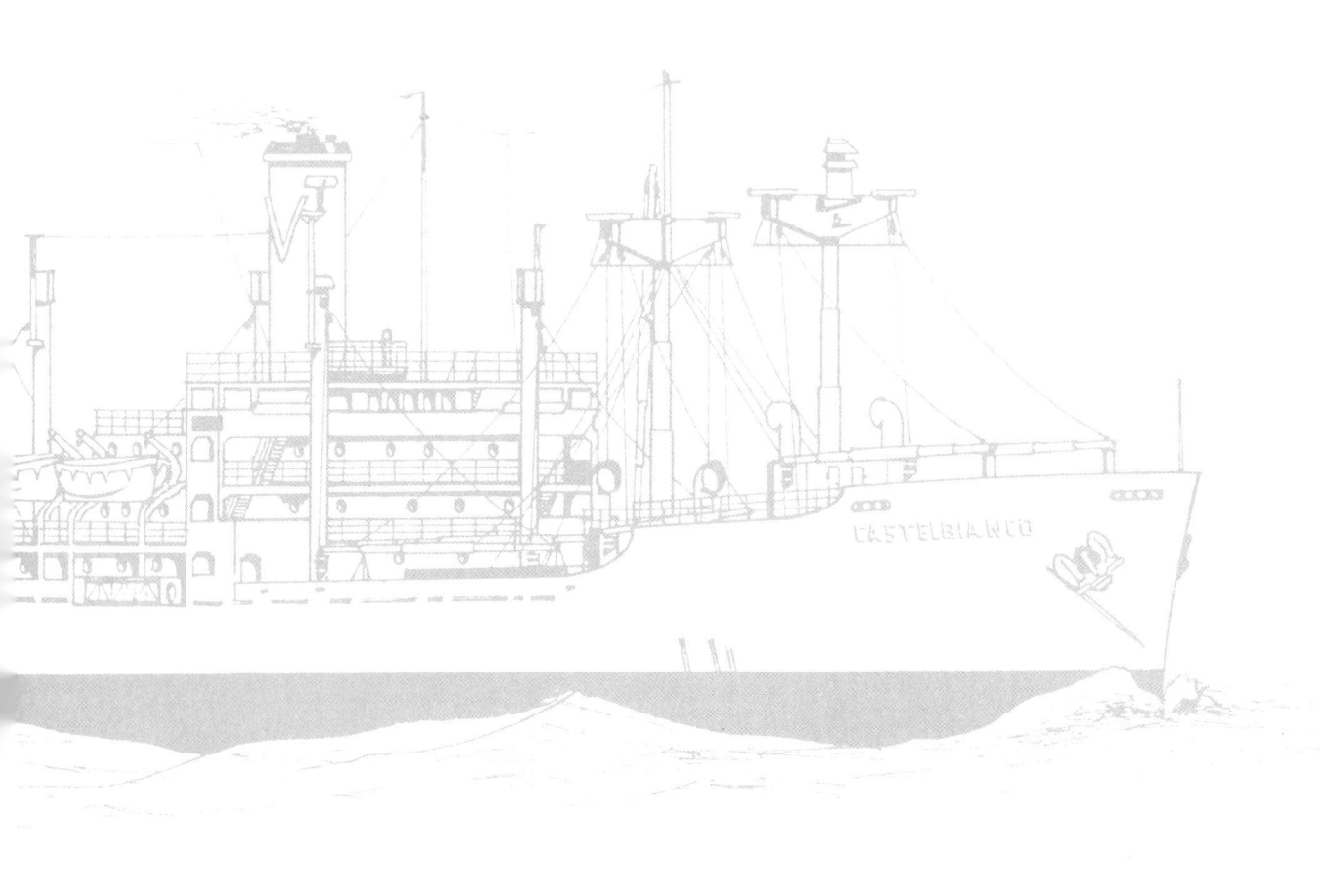

Index

Foreword	5
Introduction & Acknowledgements	6
Chronology	11
Freighters and Bulk Carriers	22
Freighters and Bulk Carriers fleet list	26
Castelbruno (1934)	44
Castelmarino (1934)	49
Castelverde (1936)	53
Castelbianco (1936)	57
Passenger Ships	60
Castel Bianco (1947)	62
Castel Verde (1947)	68
Fairsea (1949)	74
Castel Nevoso (1949)	82
Castel Felice (1952)	87
Fairsky (1958)	96
Fairstar (1964)	103
Fairsea (1971)	113
Fairwind (1972) - Albatros (1993)	123
Fairsky (1979)	134
Fairsky (1984)	137
Sitmar FairMajesty (1989)	141
Crown Princess (1990)	145
Minerva (1996)	150
Seven Seas Navigator (1999)	155
Tankers, Ore/oil and Product Carriers	160
Tankers, Ore/oil and Product Carriers fleet list	164
Silver Line	177
Ferries of the Amar Line	190
V.Ships Today & Tomorrow	194
Emigrant Liners deck plans	204
Drawings	217
Bibliography	237
Index	239

The Sitmar Liners & the V Ships
ISBN 0 9534291-0-5

©1998 by Maurizio Eliseo
©1998 Drawings by Enrico Repetto

If not otherwise credited, photographs, drawings and other pictures
are from the Vlasov Group Archives, the Vlasov family or
the author's collection.

Art Director: Francesco Dell'Olio
GDL Comunicazione - Genova (italy)
www.gdl.it
Printed by Ferrari Grafiche - Clusone (BG) Italy
January 1999

Published by
Carmania Press
Unit 202, Station House
49, Greenwich High Road

London SE10 8JL

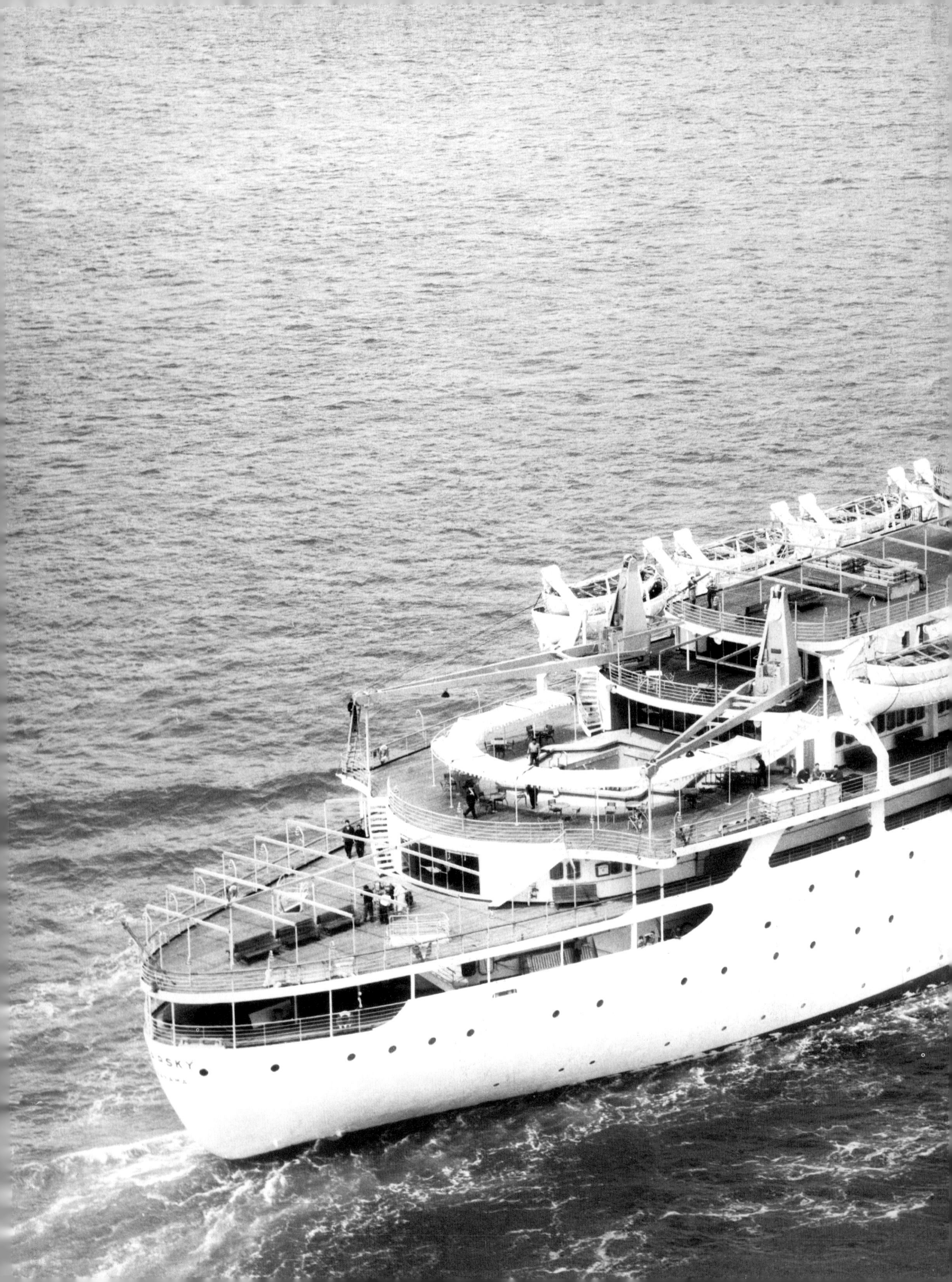